Selected Solutions Manual

C. Alton Hassell
Baylor University

General Chemistry

John W. Hill

Ralph H. Petrucci

SELECTED SOLUTIONS MANUAL

C. ALTON HASSELL
BAYLOR UNIVERSITY

GENERAL CHEMISTRY

JOHN W. HILL

RALPH H. PETRUCCI

PRENTICE HALL Upper Saddle River, NJ 07458

Production Editor: *AnnMarie Longobardo*
Production Supervisor: *Joan Eurell*
Acquisition Editor: *Mary Hornby*
Production Coordinator: *Julia Meehan*

Simon & Schuster / A Viacom Company
Upper Saddle River, NJ 07458

Printed in the United States of America

10 9 8 7 6 5 4 3 2

ISBN 0-13-569328-4

Prentice-Hall International (UK) Limited, *London*
Prentice-Hall of Australia Pty. Limited, *Sydney*
Prentice-Hall Canada, Inc., *Toronto*
Prentice-Hall Hispanoamericana, S.A., *Mexico*
Prentice-Hall of India Private Limited, *New Delhi*
Prentice-Hall of Japan, Inc., *Tokyo*
Simon & Schuster Asia Pte. Ltd., *Singapore*
Editora Prentice-Hall do Brasil, Ltda., *Rio de Janeiro*

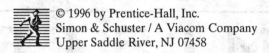

To the Student

This manual is meant to be an aid in your learning chemistry, especially to your learning to solve chemical problems. It can be misused and not be helpful, but used correctly, the manual can be a great help.

Contained in these pages are the worked out solutions for the in-chapter exercises, selected end-of-chapter review questions and the odd-numbered problems for General Chemistry by John W. Hill and Ralph H. Petrucci, Prentice Hall (1996). In working the mathematical problems, values of constants, such as the universal gas constant or the atomic weights, were used that included one more significant figure than the least well known input data. The reason was to insure that the input data would be the limiting factor in the accuracy of the answer. When the answer to an intermediate step was listed, it was usually written with one more significant figure than should be listed in the final answer. Again, this was done so that the input data, not the intermediate answer, should be the limiting value. The final answer was written with the correct number of significant figures.

Learning problem solving is much like learning to play a sport or a musical instrument. You must do it yourself; you cannot learn by watching. Skill is developed by practice and more practice. Practice time should be planned so that you are at your best. Very little will be gained by practice time when you are very tired or distracted.

Before attempting the problems, read the text chapter carefully, reading the example problems and working out the exercises. If you work on an exercise for ten or fifteen minutes and still do not have an answer, only then should you look in the solutions manual to get an idea of how to start working the problem. After the text is read, work through the review questions and the odd numbered problems. Working a problem may require looking back in the text to find a similar example. Only after ten or fifteen minutes of attempting a problem without result should the solutions manual be consulted. Any problem that requires use of the solutions manual should be reworked a few days later to see if the problem solving skill was really acquired or just the answer read from the manual.

When you refer to the manual, the solution method may be different from yours as there is usually more than one way to work a problem. There are some problems in the manual that actually show two different methods. Sometimes these solutions differ in the last significant figure because of the roundoff of numbers. If your answer is within 1 or 2 in the last significant figure, then it is probably the same number.

Don't work these problems just to get an answer. Work these problems to develop the skills to be able to solve other problems, such as those on the next test or those in your future job. While practicing these skills, it is better if you work in a quiet, undisturbed atmosphere while you are as fresh as possible. It is also better to work some every day than to cram all of your studying into one day.

Baylor University
Waco, Texas

C. Alton Hassell

Dedication

I dedicate this manual to the memory of my father, Clinton A. "Brit" Hassell (1914-1995). My greatest inheritance was the education that he provided for me.

Table of Contents

Acknowledgements

This manual is the compilation of the work of many people. I owe a great debt to each one. They have combined to keyboard, proof, standardize, edit, double-check, clarify and even beautify the initial rough draft. The errors that remain are my sole responsibility.

Adonna Cook keyboarded the manuscript. That is, she turned illegible scrawling and crude drawings into printed text and illustrative figures. Lee Ann Marshall more than once gave us the expert guidance that was needed in the use of Word.

Proofreaders and/or accuracy-checkers included Michael J. Sanger, Mark Lynch, Angela Lair, Paula Marshall, Anina Carter, Carla MacMullen, and Denise Magnuson. These wonderful, hard-working people have poured over the manual and found my many mistakes.

The Department of Chemistry at Baylor University and especially the department chairman, Dr. Marianna Busch, have given me great support and encouragement.

John Hill and Ralph Petrucci wrote a wonderful text without which this manual would be unnecessary. Ralph went the extra mile (miles) to proof and edit the manual. Robert Wismer wrote the solutions for the Special Topics.

To work with the editors and marvelous staff at Prentice Hall is to work with the best. Paul Corey signed me to the project, Ben Roberts maintained communication between the text authors and myself, Alan MacDonell helped with communication and did masterful copy editing and Mary Hornby was the glue that held us all together.

My wife, Patricia, and my children, Clint and Sharina, put up with me or with my absence during the entire process.

Chapter 1

Chemistry

Exercises:

1.1 a. physical b. chemical c. physical

1.2 a. $7.42 \times 10^{-3}\ \text{s} \times \frac{\text{ms}}{10^{-3}\ \text{s}} = 7.42\ \text{ms}$

b. $5.41 \times 10^{-6}\ \text{m} \times \frac{\mu\text{m}}{10^{-6}\text{m}} = 5.41\ \mu\text{m}$

c. $1.19 \times 10^{-3}\ \text{g} \times \frac{\text{mg}}{10^{-3}\ \text{g}} = 1.19\ \text{mg}$

d. $5.98 \times 10^{3}\ \text{m} \times \frac{\text{km}}{10^{3}\ \text{m}} = 5.98\ \text{km}$

1.3 a. $t_F = (1.8 \times 85.0\ ^\circ\text{C}) + 32 = 185\ ^\circ\text{F}$

b. $t_F = (1.8 \times -12.2\ ^\circ\text{C}) + 32.0 = 10.0\ ^\circ\text{F}$

c. $t_C = (355\ ^\circ\text{F} - 32)/1.8 = 179\ ^\circ\text{C}$

d. $t_C = (-20.8\ ^\circ\text{F} - 32.0)/1.8 = -29.3\ ^\circ\text{C}$

1.4 a. $73\ \text{m} \times 1.340\ \text{m} \times 0.41\ \text{m} = 40.1062\ \text{m}^3$ rounds to $40\ \text{m}^3$

b. $0.137\ \text{cm} \times 1.43\ \text{cm} = 0.19591\ \text{cm}^2$ rounds to $0.196\ \text{cm}^2$

c. $3.132\ \text{cm} \times 5.4\ \text{cm} \times 5.4\ \text{cm} = 91.32912\ \text{cm}^3$ rounds to $91\ \text{cm}^3$

d. $\frac{51.79\ \text{m}}{4.6\ \text{s}} = 11.258696\ \text{m/s}$ rounds to 11 m/s

e. $\frac{456.1\ \text{mi}}{7.13\ \text{h}} = 63.969144\ \text{mi/h}$ rounds to 64.0 mi/h

f. $\frac{305.5\ \text{mi}}{14.7\ \text{gal}} = 20.782313\ \text{mi/gal}$ rounds to 20.8 mi/gal

1.5 a.

$$\begin{array}{r} 48.2\ \text{m} \\ 3.82\ \text{m} \\ \underline{48.4394\ \text{m}} \\ 100.4594\ \text{m} \end{array}$$

rounds to 100.5 m

b.

$$\begin{array}{r} 148\ \ \ \ \ \ \text{g} \\ 2.39\ \ \ \text{g} \\ \underline{0.0124\ \text{g}} \\ 150.4024\ \text{g} \end{array}$$

rounds to 1.50×10^2 g

c.

$$\begin{array}{r} 451\ \ \ \ \ \text{g} \\ -\ 15.46\ \text{g} \\ \underline{-\ 20.3\ \ \ \text{g}} \\ 415.24\ \text{g} \end{array}$$

rounds to 415 g

d.
$$\begin{array}{r} 15.436\ \text{L} \\ 5.3\ \ \ \text{L} \\ -\ 6.24\ \text{L} \\ \underline{-\ 8.177\ \text{L}} \\ 6.319\ \text{L} \end{array}\ \text{rounds to } 6.3\ \text{L}$$

1.6 a. $76.3\ \text{mm} \times \frac{10^{-3}\ \text{m}}{\text{mm}} = 0.0763\ \text{m}$

b. $0.0856\ \text{kg} \times \frac{10^{3}\ \text{g}}{\text{kg}} = 85.6\ \text{g}$

c. $0.927\ \text{lb} \times \frac{16\ \text{oz}}{\text{lb}} = 14.8\ \text{oz}$

d. $415\ \text{in} \times \frac{\text{ft}}{12\ \text{in}} \times \frac{\text{yd}}{3\ \text{ft}} = 11.5\ \text{yd}$

e. $3.00\ \text{L} \times \frac{\text{mL}}{10^{-3}\ \text{L}} \times \frac{1\ \text{fl oz}}{29.57\ \text{mL}} = 101\ \text{fl oz}$

1.7 a. $\frac{90.0\ \text{km}}{\text{h}} \times \frac{\text{h}}{60\ \text{min}} \times \frac{\text{min}}{60\ \text{s}} \times \frac{10^{3}\ \text{m}}{\text{km}} = \frac{25.0\ \text{m}}{\text{s}}$

b. $\frac{1.39\ \text{ft}}{\text{s}} \times \frac{60\ \text{s}}{\text{min}} \times \frac{60\ \text{min}}{\text{h}} \times \frac{12\ \text{in}}{\text{ft}} \times \frac{2.54\ \text{cm}}{\text{in}} \times \frac{10^{-2}\ \text{m}}{\text{cm}} \times \frac{\text{km}}{10^{3}\ \text{m}} = \frac{1.53\ \text{km}}{\text{h}}$

c. $\frac{4.17\ \text{g}}{\text{s}} \times \frac{\text{kg}}{10^{3}\ \text{g}} \times \frac{60\ \text{s}}{\text{min}} \times \frac{60\ \text{min}}{\text{h}} = \frac{15.0\ \text{kg}}{\text{h}}$

1.8 a. $476\ \text{cm}^2 \times \left(\frac{1\ \text{in}}{2.54\ \text{cm}}\right)^2 = 73.8\ \text{in}^2$

b. $124\ \text{ft}^3 \times \left(\frac{12\ \text{in}}{\text{ft}}\right)^3 \times \left(\frac{2.54\ \text{cm}}{1\ \text{in}}\right)^3 \times \left(\frac{10^{-2}\ \text{m}}{\text{cm}}\right)^3 = 3.51\ \text{m}^3$

c. $\frac{15.8\ \text{lb}}{\text{in}^2} \times \left(\frac{1\ \text{in}}{2.54\ \text{cm}}\right)^2 \times \left(\frac{\text{cm}}{10^{-2}\ \text{m}}\right)^2 \times \frac{453.6\ \text{g}}{\text{lb}} \times \frac{\text{kg}}{10^{3}\ \text{g}} = \frac{1.11 \times 10^{4}\ \text{kg}}{\text{m}^2}$

1.9 $d = \frac{m}{V} = \frac{76.0\ \text{lb}}{2.54\ \text{L}} \times \frac{453.6\ \text{g}}{\text{lb}} \times \frac{10^{-3}\ \text{L}}{\text{mL}} = 13.6\ \text{g/mL}$

1.10 $V = \frac{m}{d} = 8.65\ \text{g} \times \frac{\text{mL}}{1.59\ \text{g}} = 5.44\ \text{mL}$

1.11 $10.00\ \text{gal} \times \frac{4\ \text{qt}}{\text{gal}} \times \frac{1\ \text{L}}{1.057\ \text{qt}} \times \frac{\text{mL}}{10^{-3}\ \text{L}} \times \frac{0.660\ \text{g}}{\text{mL}} \times \frac{\text{mL}}{0.791\ \text{g}} = 3.16 \times 10^{4}\ \text{mL}$

Estimation Exercises:

1.1 $200\ \text{lb} \times \frac{454\ \text{g}}{\text{lb}} \times \frac{\text{kg}}{1000\ \text{g}}$ simplifies to $200\ \text{lb} \times \frac{0.5\ \text{kg}}{\text{lb}} = 100\ \text{kg}$, so the 70 kg is a reasonable answer.

1.2 Water is about 1 g/mL

$$20 \text{ qt} \times \frac{\text{L}}{1.06 \text{ qt}} \times \frac{\text{mL}}{10^{-3} \text{ L}} \times \frac{1 \text{ g}}{\text{mL}} \times \frac{\text{kg}}{10^{3} \text{ g}} \text{ simplifies to } 20 \text{ qt} \times \frac{\text{L}}{1.1 \text{ qt}} \times \frac{\text{kg}}{\text{L}}$$

$\frac{20}{1.1}$ is closer to 20 than 15, so 20 kg is a reasonable answer.

Conceptual Exercises

1.1 The assumption that the hole is plugged was made implicitly in Conceptual Example 1.1 to obtain the density of 1.15 g/cm^3 and to conclude that the block would not float.

1.2 $V_{\text{Cyl}} = \pi r^2 h = 3.142 \times (2.5 \text{ cm})^2 \times 11.5 \text{ cm} = 226 \text{ cm}^3$

$$\text{actual density} = \frac{m}{V} = \frac{5.15 \text{ kg} \times 1000 \text{ g/kg}}{4.48 \times 10^3 \text{ cm}^3 - 226 \text{ cm}^3} = 1.21 \text{ g/cm}^3$$

mass removed = $3 \times 1.21 \text{ g/cm}^3 \times 226 \text{ cm}^3 = 820 \text{ g}$

$$5.15 \text{ kg} \times \frac{10^3 \text{ g}}{\text{kg}} = 5150 \text{ g} - 820 \text{ g} = 4.33 \times 10^3 \text{ g}$$

$$\frac{4.33 \times 10^3 \text{ g}}{4.48 \times 10^3 \text{ cm}^3} = 0.97 \text{ g/cm}^3$$

The box is less dense than water and will float.

Review Questions

13. a. physical b. chemical c. chemical d. chemical e. physical

14. C, Cl and Na are symbols representing elements. CO, $CaCl_2$ and KI are combinations of symbols for two elements (formulas) and represent compounds.

15. Helium and salt are substances. Maple syrup and vinegar are mixtures. In this case, both are water solutions.

16. Gasoline and white wine are homogeneous mixtures. Salad dressing and iced tea are heterogeneous mixtures.

17.

mass:	kilogram	(kg)
length:	meter	(m)
temperature:	kelvin	(K)
time:	second	(s)

18. area: m^2; volume: m^3; density: kg/m^3

Problems:

23. a. $4.54 \times 10^{-9} \text{ g} \times \frac{1 \text{ ng}}{10^{-9} \text{ g}} = 4.54 \text{ ng}$

b. $3.76 \times 10^{3} \text{ m} \times \frac{1 \text{ km}}{10^{3} \text{ m}} = 3.76 \text{ km}$

c. $6.34 \times 10^{-6} \text{ g} \times \frac{1 \text{ µg}}{10^{-6} \text{ g}} = 6.34 \text{ µg}$

25. a. $t_F = 1.8\ (23.5\ ^\circ C) + 32.0 = 74.3\ ^\circ F$

b. $t_C = \frac{98.6\ ^\circ F - 32.0}{1.8} = 37.0\ ^\circ C$

c. $t_F = 1.8\ (212\ ^\circ C) + 32 = 414\ ^\circ F$

27. a. $t_C = \frac{136\ ^\circ F - 32}{1.8} = 57.8\ ^\circ C$

b. $t_F = 1.8\ (-120\ ^\circ C) + 32 = -184\ ^\circ F$

29. a. $50.0\ \text{km} \times \frac{10^3\ \text{m}}{\text{km}} = 5.00 \times 10^4\ \text{m}$

b. $546\ \text{mm} \times \frac{10^{-3}\ \text{m}}{\text{mm}} = 0.546\ \text{m}$

c. $98.5\ \text{kg} \times \frac{10^3\ \text{g}}{\text{kg}} = 9.85 \times 10^4\ \text{g}$

d. $47.9\ \text{mL} \times \frac{10^{-3}\ \text{L}}{\text{mL}} = 4.79 \times 10^{-2}\ \text{L}$

e. $578\ \mu\text{s} \times \frac{10^{-6}\ \text{s}}{\mu\text{s}} \times \frac{\text{ms}}{10^{-3}\text{s}} = 0.578\ \text{ms}$

f. $237\ \text{mm} \times \frac{10^{-3}\ \text{m}}{\text{mm}} \times \frac{\text{cm}}{10^{-2}\ \text{m}} = 23.7\ \text{cm}$

31. The yardstick (3 ft.) is shorter than the 3-ft. 5-in. rattlesnake. The rattlesnake is 41 inches [(3 x 12) + 5], which is shorter than the chain, because one meter is about 40 inches, so the chain is about 48 (40 x 1.2) inches. The rope is longest at 75 in.

$(4) < (3) < (1) < (2)$

33. a. 4 b. 2 c. 5 d. 3 e. 4 f. 4

35. a. 2.800×10^3 m b. 9.000×10^3 s

c. 9.0×10^{-4} cm d. 2.000×10^1 s

37. a. 48.2 m
3.82 m
48.4394 m
100.4594 =>100.5 m

b. 151 g
2.39 g
.0124 g
153.4024 => 153 g

c. 100.53 cm
- 46.1 cm
54.43 => 54.4 cm

d. 451 g
- 15.46 g
435.54 => 436 g

e. 15.44 mL
- 9.1
105
111.34 => 111 mL

f. 12.52 cm
+ 5.1 cm
- 3.18 cm
-12.02 cm
2.42 => 2.4 cm

39. a. $73.0 \times 1.340 \times 0.41 = 40.1062 \Rightarrow 40$

b. $265.02 \times 0.000581 \times 12.18 = 1.8754352 \Rightarrow 1.88$

c. $\dfrac{33.58 \times 1.007}{0.00705} = 4796.4624 \Rightarrow 4.80 \times 10^3$

d. $\dfrac{22.61 \times 0.0587}{135 \times 28} = 0.000351113 \Rightarrow 3.5 \times 10^{-4}$

41. $d = \dfrac{m}{V} = \dfrac{78.0 \text{ g}}{25.0 \text{ mL}} = 3.12 \text{ g/mL}$

43. 18.43 g metal + paper
- 1.2140 g paper
17.22 metal

$d = \dfrac{m}{V} = \dfrac{17.22 \text{ g}}{3.29 \text{ cm}^3} = 5.23 \text{ g/cm}^3$

45. $d = \dfrac{m}{V} = \dfrac{59.01 \text{ g}}{1.20 \text{ cm} \times 2.41 \text{ cm} \times 1.80 \text{ cm}} = \dfrac{59.01 \text{ g}}{5.206 \text{ cm}^3} = 11.3 \text{ g/cm}^3$

47. $m = dV = 30.0 \text{ mL} \times \dfrac{1.32 \text{ g}}{\text{mL}} = 39.6 \text{ g}$

49. $V = \dfrac{m}{d} = 898 \text{ kg} \times \dfrac{10^3 \text{ g}}{\text{kg}} \times \dfrac{\text{cm}^3}{7.76 \text{ g}} = 1.16 \times 10^5 \text{ cm}^3$

$l = \dfrac{V}{A} = \dfrac{1.16 \times 10^5 \text{ cm}^3}{1.50 \text{ cm}^2} = 7.73 \times 10^4 \text{ cm}$

51. $d = \dfrac{m}{V} = \dfrac{3.2 \text{ kg}}{0.80 \text{ m} \times 0.80 \text{ m} \times 1.20 \text{ m}} \times \dfrac{10^3 \text{ g}}{\text{kg}} \times \left(\dfrac{10^{-2} \text{ m}}{\text{cm}}\right)^3 = \dfrac{3.2 \times 10^3 \text{ g}}{7.68 \times 10^5 \text{ cm}^3}$

$= 4.2 \times 10^{-3} \text{ g/cm}^3$

53. $15 \text{ gal} \times \frac{4 \text{ qt}}{\text{gal}} \times \frac{\text{L}}{1.06 \text{ qt}} \times \frac{\text{mL}}{10^{-3} \text{ L}} \times \frac{1 \text{ g}}{\text{mL}} \times \frac{\text{lb}}{454 \text{ g}}$

estimate $\frac{15 \times 4 \times 1000}{1 \times 500} \approx 120$ lb (actually 125 lb) water

$3.0 \text{ L} \times \frac{\text{mL}}{10^{-3} \text{ L}} \times \frac{\text{cm}^3}{\text{mL}} \times \frac{13.6 \text{ g}}{\text{cm}^3} \times \frac{\text{lb}}{454 \text{ g}} =$

estimate $\frac{3 \times 1000 \times 14}{500} = 84$ lb (actually 90 lb) mercury

(2) The water is most difficult to lift.

55. The meter stick, with 1000 mm markings, is actually 1005 mm long. The room is longer by 5 mm for every meter measured. The true room length is 14.155 m x $\frac{1.005 \text{ m}}{1.000 \text{ m}} = 14.226$ m. The error in the measured length is 14.226 - 14.155 = 0.071 m.

57. $1 \text{ link} \times \frac{\text{chain}}{100 \text{ links}} \times \frac{\text{furlong}}{10 \text{ chain}} \times \frac{\text{mi}}{8 \text{ furlongs}} \times \frac{5280 \text{ ft}}{\text{mi}} \times \frac{12 \text{ in}}{\text{ft}} = 7.92 \text{ in}$

59. $\frac{0.998 \text{ g}}{\text{cm}^3} \times \frac{\text{lb}}{453.6 \text{ g}} \times \left(\frac{2.54 \text{ cm}}{\text{in}}\right)^3 \times \left(\frac{12 \text{ in}}{\text{ft}}\right)^3 = \frac{62.3 \text{ lb}}{\text{ft}^3}$

61.

4.72 kg	bottle and wine
1.70 kg	bottle
3.02 kg	wine

$d = \frac{3.02 \text{ kg}}{3.00 \text{ L}} \times \frac{10^3 \text{ g}}{\text{kg}} \times \frac{10^{-3} \text{ L}}{\text{mL}} = 1.007 \text{ g/mL}$

$275 \text{ mL} \times \frac{1.007 \text{ g wine}}{\text{mL}} \times \frac{11.0 \text{ g alcohol}}{100 \text{ g wine}} \times \frac{\text{oz}}{28.35 \text{ g}} = 1.07 \text{ oz}$

63. $275 \text{ g} \times \frac{470 \text{ m}^2}{\text{g}} = 1.3 \times 10^5 \text{ m}^2$

65. $$d = \frac{5.15 \text{ kg}}{30.2 \text{ cm x } 12.9 \text{ cm x } 11.5 \text{ cm - } \pi\,(2.5 \text{ cm})^2\; 11.5 \text{ cm}} = \frac{5.15 \text{ kg}}{4.48 \times 10^3 \text{ cm}^3 - 0.23 \times 10^3 \text{ cm}^3}$$

$$d = \frac{5.15 \text{ kg}}{4.25 \times 10^3 \text{ cm}^3} \times \frac{1000\text{g}}{\text{kg}} = 1.21 \text{ g/cm}^3$$ density of material in the block

density of water = 1.0 g/cm^3

Volume of box = 30.2 cm x 12.9 cm x 11.5 cm = 4.48 x 10^3 cm^3

4.48 x 10^3 cm^3 x 1.00 g/cm^3 = 4480 g
This is the mass of box that would be equal to the mass of an equal volume of water.

5150 g - 4480 g = 670 g. This is the mass of material to be cut out.

$670 \text{ g x} \left(\frac{\text{cm}^3}{1.21 \text{ g}}\right) = 554 \text{ cm}^3$. This is the volume of material to be cut out.

554 cm^3 + 226 cm^3 = 780 cm^3. This is the volume of the hole.

$\frac{780 \text{ cm}^3}{11.5 \text{ cm}} = 67.8 \text{ cm}^2$. This is the area of the hole.

$\sqrt{\frac{67.8 \text{ cm}^2}{\pi}} = 4.65 \text{ cm}$. This is the radius of the hole.

diameter = 9.30 cm

The hole must be larger for any of the box to be above the water line. When the box is the density of water it will float in water, but none of the box will be above the top of the water. Also, the ends of the hole must be plugged with something and this will reintroduce some mass to the box.

67. $V_{\text{brass}} = (2.0 \text{ cm})^3 = 8.0 \text{ cm}^3$
The brass cube will sink to the bottom displacing 8.0 cm^3 or 8.0 mL of water.
$V_{\text{cork}} = 5.0 \text{ cm x } 4.0 \text{ cm x } 2.0 \text{ cm} = 40 \text{ cm}^3$
$m_{\text{cork}} = dV = \frac{0.22 \text{ g}}{\text{cm}^3} \text{ x } 40 \text{ cm}^3 = 8.8 \text{ g}$
The cork will displace 8.8 g of water, which is about 8.8 mL of water. More water will overflow from the vessel of water in which the cork is floated (right).

Chapter 2

Laws, Symbols, and Formulas

Exercises

2.1   a. $0.612 : 1.142 = 0.536 : 1$ not possible
       b. $1.250 : 1.142 = 1.095 : 1$ not possible
       c. $1.713 : 1.142 = 1.5 : 1$ or $3 : 2$ possible
       d. $2.856 : 1.142 = 2.5 : 1$ or $5 : 2$ possible

2.2 $A = Z + \text{\# neutrons} = 50 + 66 = 116$

2.3   tetrasulfur dinitride

2.4 P_4O_{10}

2.5   a. AlF_3 b. K_2S c. Ca_3N_2 d. Li_2O

2.6   a. calcium bromide
       b. lithium sulfide
       c. iron(II) bromide
       d. copper(I) iodide

2.7   a. NH_4NO_3
       b. iron(III) phosphate
       c. NaClO
       d. potassium hydrogen carbonate

2.8 C - C - C - C - C - C

```
                                      C
                                      |
                              C - C - C - C - C
```

```
      C
      |
C - C - C - C - C
```

```
          C
          |
          C
          |
  C - C - C - C
          |
          C
```

```
  C   C
  |   |
C - C - C - C
```

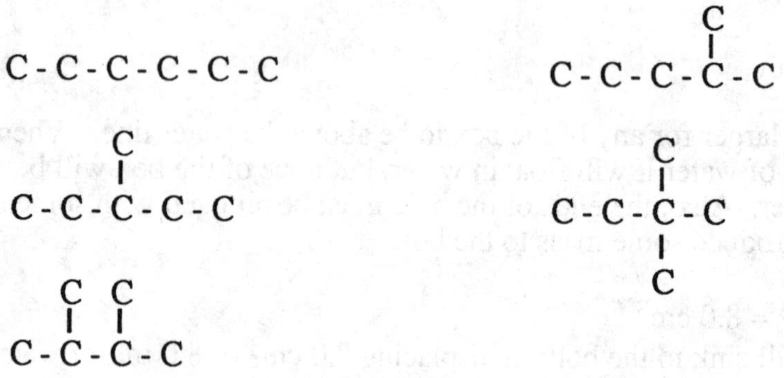

Estimation Exercise

2.1   Octane has two more CH_2 groups than hexane, so its boiling point is about 69 °C + (2 x 30°) = 129 °C.

Conceptual Exercise

2.1 a. No, the formulas are different: C_7H_{16} for the left structure and C_7H_{14} for the cyclic structure on the right.
b. Yes, both hydrocarbons have the formula C_9H_{20}. One has methyl groups on the second and fourth carbon atoms, and the other, on the second and fifth carbon atoms.

Review Questions

4. Law of definite proportions.

5. a. The total mass is 53.00 g.
b. 11.00 g of carbon dioxide is formed.
c. Law of conservation of mass and law of definite proportions.

6. Compound B has an oxygen-to-sulfur mass ratio of 3:2. The law of multiple proportions is illustrated.

7. The law of conservation of mass.

17. a. different elements; b. isotopes; c. identical atoms; d. different elements; e. different elements.

18. a. ${}^{8}_{5}B$ b. ${}^{14}_{6}C$ c. ${}^{235}_{92}U$ d. ${}^{60}_{27}Co$

23. a. Magnesium is an element.
b. Hydroxyl is a prefix, meaning an OH group.
c. Chloride is a negatively charged atom of chlorine, the ion Cl^-.
d. Ammonia is a compound, NH_3.
e. Ammonium is an ion, NH_4^+.
f. Ethane is a compound, CH_3CH_3.
The only substances that could be found on a stockroom shelf are (a), (d), and (f).

29. a. Usually it is organic if it contains C, but there are a few exceptions, so a molecular formula is sufficient.
b. It contains only H and C, so a molecular formula is sufficient.
c. An alcohol requires a structural formula.
d. It contains only H and C and the ratio is $2n + 2$ hydrogens for each carbon, so a molecular formula is sufficient.
e. A carboxylic acid requires a structural formula.

Problems

31. $Zn + S \longrightarrow ZnS$
Before reaction: 1.000 g Zn + 0.200 g S = 1.200 g
After reaction: 0.608 g ZnS + 0.592 g Zn = 1.200 g
The mass of substances after the reaction equals the mass of substances before the reaction. The law of conservation of mass is confirmed.

33. $\frac{0.625\ g}{1.000\ g}\ 100\% = 62.5\%\ C$ $\quad \frac{0.0419\ g}{1.000g}\ 100\% = 4.19\%\ H$

$\frac{0.968\ g}{1.549\ g}\ 100\% = 62.5\%\ C$ $\quad \frac{0.0649\ g}{1.549\ g}\ 100\% = 4.19\%\ H$

$\frac{0.618\ g}{0.988\ g}\ 100\% = 62.6\%\ C$ $\quad \frac{0.0414\ g}{0.988\ g}\ 100\% = 4.19\%\ H$

Yes, each sample has the same percent carbon and percent hydrogen.

35. A and D are isotopes; B and C are isotopes

37.

		Electrons	Protons
a.	Ca	20	20
b.	Na	11	11
c.	F	9	9
d.	Ar	18	18
e.	Be	4	4

39.

		Protons	Neutrons
a.	^{62}Zn	30	32
b.	^{241}Pu	94	147
c.	^{99}Tc	43	56
d.	^{99}Mo	42	57

41.

		Group	Period	Type
a.	C	4A	2	nonmetal
b.	Ca	2A	4	metal
c.	Cd	2B	5	metal
d.	Cl	7A	3	nonmetal
e.	B	3A	2	nonmetal
f.	Ba	2A	6	metal
g.	Bi	5A	6	metal
h.	Br	7A	4	nonmetal

43. a. gallium
b. copper
c. iodine

45. a. He b. O_2 c. Cl_2 d. P_4

47. a. N_2O b. P_4S_3 c. PCl_5 d. SF_6

49. a. carbon disulfide
b. dinitrogen tetrasulfide
c. phosphorus trifluoride
d. disulfur decafluoride

51. a. sodium ion b. magnesium ion
c. aluminum ion d. chloride ion
e. oxide ion f. nitride ion

53. a.   iron(III) or ferric ion
     b.   copper(II) or cupric ion
     c.   silver ion

55. a. Br^- b. Ca^{2+} c. K^+ d. Fe^{2+} e. Na^+

57. a.   carbonate ion
     b.   hydrogen phosphate ion
     c.   permanganate ion
     d.   hydroxide ion

59. a. NH_4^+ b. HSO_4^- c. CN^- d. NO_2^-

61. a.   sodium bromide   b.   iron(III) chloride
     c.   lithium iodide   d.   sodium oxide
     e.   potassium sulfide   f.   copper(I) bromide
     g.   potassium chloride   h.   magnesium bromide
     i.   calcium sulfide   j.   iron(II) chloride
     k.   aluminum oxide

63. a. $MgSO_4$ b. $NaHCO_3$ c. KNO_3 d. $CaHPO_4$ e. $Ca(ClO_2)_2$
     f. $CaCO_3$ g. $LiHSO_4$ h. $Mg(CN)_2$ i. KH_2PO_4 j. $NaClO$

65. a.   sodium hydrogen sulfate
     b.   aluminum hydroxide
     c.   sodium carbonate
     d.   potassium hydrogen carbonate
     e.   ammonium nitrite

67. a. HCl b. H_2SO_4 c. H_2CO_3 d. HCN e. $LiOH$ f. $Mg(OH)_2$

69. a.   sodium hydroxide   b.   phosphoric acid
     c.   nitric acid   d.   sulfurous acid
     e.   calcium hydroxide   f.   hydrosulfuric acid

71.

```
        H H H H H                    H H H O                    H H H
        | | | | |                    | | | ||                   | | |
   a.  H-C-C-C-C-C-H            b.  H-C-C-C-C-OH           c.  H-C-C-C-OH
        | | | | |                    | | |                      | | |
        H H H H H                    H H H                      H H H


                                    H H H
           H H                       | | |
           | |              e.  H-C-C-C-H
   d.   H-C-C-H                      H | H                         H O
           | |                         |                          | ||
        H-C-C-H                     H-C-H                  f.  H-C-C-OH
           | |                         |                          |
           H H                         H                          H
```

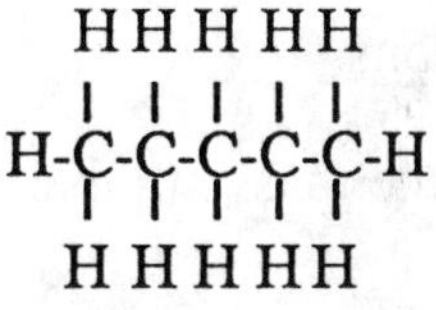

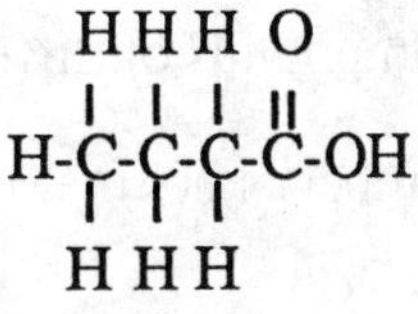

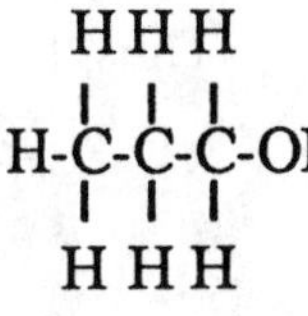

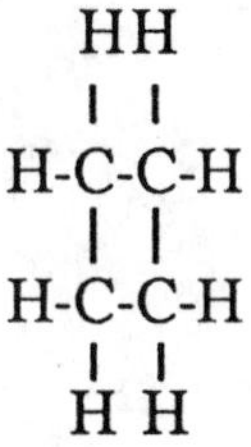

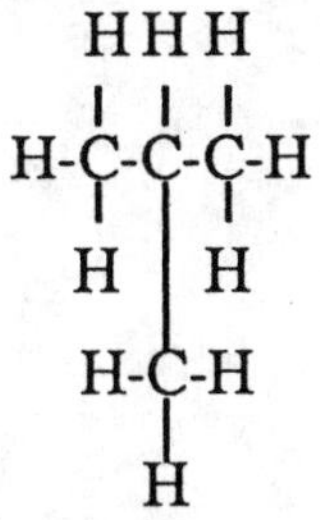

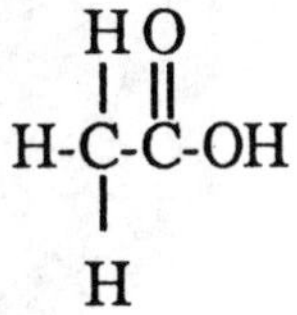

73. $CH_3CH_2CH_2\overset{O}{\overset{||}{C}}OH$ butyric acid $CH_3CH(CH_3)\overset{O}{\overset{||}{C}}OH$ isobutyric acid

75. a. straight-chain alkane hydrocarbon
b. alcohol
c. hydrocarbon (could be cyclic alkane)
d. hydrocarbon
e. carboxylic acid
f. inorganic compound

77. b: 2,2,4-trimethylpentane

79. Before reaction:

$$100.0 \text{ mL x } \frac{1.148\text{g}}{\text{mL}} = 114.8 \text{ g HCl solution}$$

114.8 g HCl solution
+10.00 g $CaCO_3$
124.8 g total

After reaction:

$$2.22 \text{ L x } \frac{0.0019769 \text{ g}}{\text{mL}} \text{ x } \frac{\text{mL}}{10^{-3} \text{ L}} = 4.39 \text{ g } CO_2$$

4.39 g CO_2 gas
+120.40 g solution
124.79 g total

The mass of substances after the reaction equals the mass of substances before the reaction. The law of conservation of mass is confirmed.

81. $$0.842 \text{ g } SO_2 \text{ x } \frac{0.312 \text{ g S}}{0.623 \text{ g } SO_2} = 0.422 \text{ g S burned}$$

83. In a 100 g sample,

$$\frac{96.2 \text{ g Hg}}{3.8 \text{ g O}} = 25:1 \qquad \frac{92.6 \text{ g Hg}}{7.4 \text{ g O}} = 12.5:1 \qquad \frac{25}{12.5} = 2:1$$

Thus, there are two compounds of different proportions.

85. The atomic number is 20, that is, calcium. The mass number is 40, so the isotope is $^{40}_{20}Ca$.

87. a. There is only one position where the OH group can be substituted on the ethanol molecule; ethanol does not have an isomer.
    b. The third carbon would have to form five bonds, but only four bonds are possible.

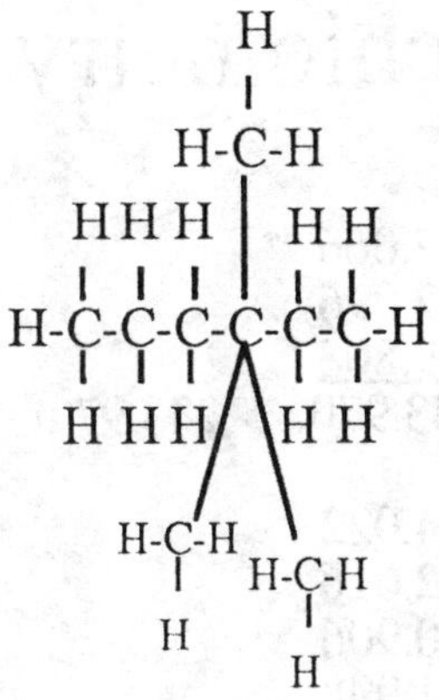

89. a. No. One structure is simply a flipped-over version of the other.
    b. Yes. The structures have the same formula (C_9H_{20}) but the positions of the CH_3 group are different.
    c. No. The structures have different formulas.
    d. No. The structures and formulas are identical.

91. a. isobutanol or 2-methyl-1-propanol
    b. 3-methyl-2-pentanol
    c. 2,2,4-trimethylpentane
    d. 2,2-dimethyl-1-propanol

Chapter 3

Stoichiometry

Exercises

3.1 a.

C	6 x 12.011 =	72.066
H	12 x 1.008 =	12.096
Br	2 x 79.904 =	159.808
		243.970 => 243.97 u

b.

C	2 x 12.011 =	24.022
H	2 x 1.008 =	2.016
Cl	2 x 35.453 =	70.906
O	2 x 15.999 =	31.998
		128.942 => 128.94 u

c.

H	5 x 1.008 =	5.040
P	3 x 30.974 =	92.922
O	10 x 15.999 =	159.990
		257.952 => 257.95 u

d.

C	6 x 12.011 =	72.066
H	4 x 1.008 =	4.032
Cl	2 x 35.453 =	70.906
O	2 x 15.999 =	31.998
S	1 x 32.066 =	32.066
		211.068 => 211.07 u

3.2 a.

K	2 x 39.098 =	78.196
Sb	1 x 121.75 =	121.75
F	5 x 18.998 =	94.990
		294.936 => 294.94 u

b. $Ba(BrO_3)_2$

Ba	1 x 137.33 =	137.33
Br	2 x 79.904 =	159.808
O	6 x 15.999 =	95.994
		393.132 => 393.13 u

c. $FePO_4$

Fe	1 x 55.847 =	55.847
P	1 x 30.974 =	30.974
O	4 x 15.999 =	63.996
		150.817 => 150.82 u

d.

Na	1 x 22.990 =	22.990
B	1 x 10.811 =	10.811
C	24 x 12.011 =	288.264
H	20 x 1.008 =	20.160
		342.225 => 342.23 u

3.3 a. $3.71 \text{ g Fe} \times \frac{\text{mol Fe}}{55.85 \text{ g Fe}} = 0.0664 \text{ mol Fe}$

b.

H	3 x 1.008 =	3.024
P	1 x 30.97 =	30.97
O	4 x 16.00 =	64.00
		97.99 u

$76.0 \text{ mg } H_3PO_4 \times \frac{10^{-3} \text{ g}}{\text{mg}} \times \frac{\text{mol } H_3PO_4}{97.99 \text{ g } H_3PO_4} = 7.76 \times 10^{-4} \text{ mol } H_3PO_4$

c. $4 \times 12.01 + 10 \times 1.008 = 58.12 \text{ u}$

$1.65 \times 10^3 \text{ kg } C_4H_{10} \times \frac{10^3 \text{ g}}{\text{kg}} \times \frac{\text{mol } C_4H_{10}}{58.12 \text{ g } C_4H_{10}} = 2.84 \times 10^4 \text{ mol } C_4H_{10}$

d. K_2CrO_4 $2 \times 39.10 + 1 \times 52.00 + 4 \times 16.00 = 194.20 \text{ u}$

$1.99 \text{ g } K_2CrO_4 \times \frac{\text{mol } K_2CrO_4}{194.20 \text{ g } K_2CrO_4} = 0.0102 \text{ mol } K_2CrO_4$

3.4 a. H_2O $2 \times 1.008 + 1 \times 16.00 = 18.02 \text{ u}$

$55.5 \text{ mol } H_2O \times \frac{18.02 \text{ g } H_2O}{\text{mol } H_2O} = 1.00 \times 10^3 \text{ g } H_2O$

b. $C_4H_{10}O$ $4 \times 12.01 + 10 \times 1.008 + 1 \times 16.00 = 74.12 \text{ u}$

$0.0102 \text{ mol } C_4H_{10}O \times \frac{74.12 \text{ g } C_4H_{10}O}{\text{mol } C_4H_{10}O} = 0.756 \text{ g } C_4H_{10}O$

c. C_2H_6 $2 \times 12.01 + 6 \times 1.008 = 30.07 \text{ u}$

$2.45 \times 10^{-4} \text{ mol } C_2H_6 \times \frac{30.07 \text{ g } C_2H_6}{\text{mol } C_2H_6} = 7.37 \times 10^{-3} \text{ g } C_2H_6$

d. HNO_3 $1 \times 1.008 + 1 \times 14.01 + 3 \times 16.00 = 63.02 \text{ u}$

$2.13 \text{ mol } HNO_3 \times \frac{63.02 \text{ g } HNO_3}{\text{mol } HNO_3} = 134 \text{ g } HNO_3$

3.5 a. $6.17 \text{ g Ca} \times \frac{\text{mol Ca}}{40.08 \text{ g Ca}} \times \frac{6.022 \times 10^{23} \text{ atom}}{\text{mol}} = 9.27 \times 10^{22} \text{ atoms}$

b. $0.0100 \text{ g } N_2 \times \frac{\text{mol } N_2}{28.01 \text{ g } N_2} \times \frac{6.022 \times 10^{23} \text{ molecules}}{\text{mol}} =$

2.15×10^{20} molecules

c. C_4H_{10} $4 \times 12.01 + 10 \times 1.008 = 58.12 \text{ u}$

$18.5 \text{ g } C_4H_{10} \times \frac{\text{mol } C_4H_{10}}{58.12 \text{ g } C_4H_{10}} \times \frac{6.022 \times 10^{23} \text{ molecules}}{\text{mol}} =$

1.92×10^{23} molecules

d. $12 + 22 + 11 = 45$ atoms/molecule

$C_{12}H_{22}O_{11}$ $12 \times 12.01 + 22 \times 1.008 + 11 \times 16.00 = 342.30$ u

$$215 \text{ g } C_{12}H_{22}O_{11} \times \frac{\text{mol } C_{12}H_{22}O_{11}}{342.30 \text{ g } C_{12}H_{22}O_{11}} \times \frac{6.022 \times 10^{23} \text{ molecules}}{\text{mol}} \times \frac{45 \text{ atoms}}{\text{molecule}} = 1.70 \times 10^{25} \text{ atoms}$$

3.6 a. $$\frac{4.003 \text{ g}}{\text{mol}} \times \frac{\text{mol}}{6.022 \times 10^{23} \text{ atoms}} = 6.65 \times 10^{-24} \text{ g}$$

b. $$\frac{208.98 \text{ g}}{\text{mol}} \times \frac{\text{mol}}{6.022 \times 10^{23} \text{ atoms}} = 3.47 \times 10^{-22} \text{ g}$$

c. CCl_4 $1 \times 12.01 + 4 \times 35.45 = 153.8$ u

$$\frac{153.8 \text{ g}}{\text{mol}} \times \frac{\text{mol}}{6.022 \times 10^{23} \text{ molecules}} = 2.55 \times 10^{-22} \text{ g}$$

d. C_3H_8 $3 \times 12.01 + 8 \times 1.008 = 44.09$ u

$$\frac{44.09 \text{ g}}{\text{mol}} \times \frac{\text{mol}}{6.022 \times 10^{23} \text{ molecules}} = 7.32 \times 10^{-23} \text{ g}$$

3.7 a. $(NH_4)_2SO_4$

$2 \times 14.01 + 8 \times 1.008 + 1 \times 32.07 + 4 \times 16.00 = 132.15$ u

$$\%N = \frac{28.02 \text{ g N}}{132.15 \text{ g } (NH_4)_2SO_4} \times 100\% = 21.20\% \text{ N}$$

$$\%H = \frac{8.064 \text{ g H}}{132.15 \text{ g } (NH_4)_2SO_4} \times 100\% = 6.10\% \text{ H}$$

$$\%S = \frac{32.07 \text{ g S}}{132.15 \text{ g } (NH_4)_2SO_4} \times 100\% = 24.27\% \text{ S}$$

$$\%O = \frac{64.00 \text{ g O}}{132.15 \text{ g } (NH_4)_2SO_4} \times 100\% = \underline{48.43\% \text{ O}}$$

100.00%

b. $CO(NH_2)_2$ $1 \times 12.01 + 1 \times 16.00 + 2 \times 14.01 + 4 \times 1.008 = 60.06$ u

$$\%N = \frac{28.02 \text{ g N}}{60.06 \text{ g } CO(NH_2)_2} \times 100\% = 46.65\% \text{ N}$$

$$\%C = \frac{12.01 \text{ g C}}{60.06 \text{ g } CO(NH_2)_2} \times 100\% = 20.00\% \text{ C}$$

$$\%O = \frac{16.00 \text{ g O}}{60.06 \text{ g } CO(NH_2)_2} \times 100\% = 26.64\% \text{ O}$$

$$\%H = \frac{4.032 \text{ g H}}{60.06 \text{ g } CO(NH_2)_2} \times 100\% = \underline{6.71\% \text{ H}}$$

100.00%

$$\%N = \frac{28.02 \text{ g N}}{2 \times 14.01 + 4 \times 1.008 + 3 \times 16.00} = 35.00\% \text{ N in } NH_4NO_3$$

Urea has the highest %N.

3.8 $NaHCO_3$ $1 \times 22.99 + 1 \times 1.008 + 1 \times 12.01 + 3 \times 16.00 = 84.01$ u

$$5.00\text{g } NaHCO_3 \times \frac{\text{mol } NaHCO_3}{84.01 \text{ g } NaHCO_3} \times \frac{\text{mol Na}}{\text{mol } NaHCO_3} \times \frac{22.99 \text{ g Na}}{\text{mol Na}} \times \frac{\text{mg}}{10^{-3}\text{ g}} =$$

1.37×10^3 mg Na

3.9

$$51.70 \text{ g C} \times \frac{\text{mol C}}{12.01 \text{ g C}} = 4.305 \text{ mol C} \times \frac{1}{0.8608 \text{ mol N}} = 5.00$$
$$8.68 \text{ g H} \times \frac{\text{mol H}}{1.008 \text{ g H}} = 8.61 \text{ mol H} \times \frac{1}{0.8608 \text{ mol N}} = 10.0$$
$$12.06 \text{ g N} \times \frac{\text{mol N}}{14.01 \text{ g N}} = 0.8608 \text{ mol N} \times \frac{1}{0.8608 \text{ mol N}} = 1.00$$
$$27.55 \text{ g O} \times \frac{\text{mol O}}{16.00 \text{ g O}} = 1.722 \text{ mol O} \times \frac{1}{0.8608 \text{ mol N}} = 2.00$$

$C_5H_{10}NO_2$

3.10

$$37.01 \text{ g C} \times \frac{\text{mol C}}{12.01 \text{ g C}} = 3.082 \text{ mol C} \times \frac{1}{1.320 \text{ mol N}} = 2.335 \times 3 = 7$$
$$2.22 \text{ g H} \times \frac{\text{mol H}}{1.008 \text{ g H}} = 2.20 \text{ mol H} \times \frac{1}{1.320 \text{ mol N}} = 1.67 \times 3 = 5$$
$$18.50 \text{ g N} \times \frac{\text{mol N}}{14.01 \text{ g N}} = 1.320 \text{ mol N} \times \frac{1}{1.320 \text{ mol N}} = 1.00 \times 3 = 3$$
$$42.27 \text{ g O} \times \frac{\text{mol O}}{16.00 \text{ g O}} = 2.642 \text{ mol O} \times \frac{1}{1.320 \text{ mol N}} = 2.00 \times 3 = 6$$

$C_7H_5N_3O_6$ (multiplication by 3 to obtain whole numbers)

3.11 CH_2 $1 \times 12.01 + 2 \times 1.008 = 14.0$ u

ethylene $$\frac{28.0 \text{ u}}{\text{molecule}} \times \frac{\text{formula}}{14.0 \text{ u}} = \frac{2 \text{ formula}}{\text{molecule}}$$

C_2H_4

cyclohexane $$\frac{84.0 \text{ u}}{\text{molecule}} \times \frac{\text{formula}}{14.0 \text{ u}} = \frac{6 \text{ formula}}{\text{molecule}}$$

C_6H_{12}

1-pentene $$\frac{70.0 \text{ u}}{\text{molecule}} \times \frac{\text{formula}}{14.0 \text{ u}} = \frac{5 \text{ formula}}{\text{molecule}}$$

C_5H_{10}

3.12 a.

$$1.067 \text{ g } CO_2 \times \frac{12.011 \text{ g C}}{44.010 \text{ g } CO_2} \times \frac{100\%}{0.3629 \text{ g sample}} = 80.24\% \text{ C}$$
$$0.3120 \text{ g } H_2O \times \frac{2 \times 1.0079 \text{ g H}}{18.015 \text{ g } H_2O} \times \frac{100\%}{0.3629 \text{ g sample}} = 9.62\% \text{ H}$$

100.00% - 80.24% C - 9.62% H = 10.14% O

b.

$$80.24 \text{ g C} \times \frac{\text{mol C}}{12.011 \text{ g C}} = 6.681 \text{ mol C} \times \frac{1}{0.6338 \text{ mol O}} = 10.54$$
$$9.62 \text{ g H} \times \frac{\text{mol H}}{1.0079 \text{ g H}} = 9.545 \text{ mol H} \times \frac{1}{0.6338 \text{ mol O}} = 15.06$$
$$10.14 \text{ g O} \times \frac{\text{mol O}}{15.999 \text{ g O}} = 0.6338 \text{ mol O} \times \frac{1}{0.6338 \text{ mol O}} = 1.00$$

$C_{10.54}H_{15.00}O_{1.00}$ —> $C_{21}H_{30}O_2$

3.13 a. $3\ Mg + B_2O_3 \longrightarrow 2\ B + 3\ MgO$
b. $3\ NO_2 + H_2O \longrightarrow 2\ HNO_3 + NO$
c. $3\ H_2 + Fe_2O_3 \longrightarrow 2\ Fe + 3\ H_2O$
d. $6\ CaO + P_4O_{10} \longrightarrow 2\ Ca_3(PO_4)_2$
e. $C_5H_{12} + 8\ O_2 \longrightarrow 5\ CO_2 + 6\ H_2O$
f. $2\ C_4H_{10} + 13\ O_2 \longrightarrow 8\ CO_2 + 10\ H_2O$

3.14 a. $C_3H_8 + 5\ O_2 \longrightarrow 3\ CO_2 + 4\ H_2O$

$$0.529 \text{ mol } C_3H_8 \times \frac{3 \text{ mol } CO_2}{\text{mol } C_3H_8} = 1.59 \text{ mol } CO_2$$

b.
$$76.2 \text{ mol } C_3H_8 \times \frac{4 \text{ mol } H_2O}{\text{mol } C_3H8} = 305 \text{ mol } H_2O$$

c.
$$1.010 \text{ mol } O_2 \times \frac{3 \text{ mol } CO_2}{5 \text{ mol } O_2} = 0.6060 \text{ mol } CO_2$$

3.15 $2\ Mg\ (s) + TiCl_4 \longrightarrow 2\ MgCl_2 + Ti$

$$83.6 \text{ g } TiCl_4 \times \frac{\text{mol } TiCl_4}{189.68 \text{ g } TiCl_4} \times \frac{2 \text{ mol Mg}}{\text{mol } TiCl_4} \times \frac{24.31 \text{ g Mg}}{\text{mol Mg}} = 21.4 \text{ g Mg}$$

3.16 $2\ KClO_3 \longrightarrow 2\ KCl + 3\ O_2$

$$2.47 \text{ g } KClO_3 \times \frac{\text{mol } KClO_3}{122.55 \text{ g } KClO_3} \times \frac{3 \text{ mol } O_2}{2 \text{ mol } KClO_3} \times \frac{32.00 \text{ g } O_2}{\text{mol } O_2} = 0.967 \text{ g } O_2$$

3.17 $2\ C_8H_{18} + 25\ O_2 \longrightarrow 16\ CO_2 + 18\ H_2O$

$$775 \text{ mL } C_8H_{18} \times \frac{0.7025 \text{ g } C_8H_{18}}{\text{mL } C_8H_{18}} \times \frac{\text{mol } C_8H_{18}}{114.22 \text{ g } C_8H_{18}} \times \frac{18 \text{ mol } H_2O}{2 \text{ mol } C_8H_{18}} \times$$

$$\frac{18.015 \text{ g } H_2O}{\text{mol } H_2O} \times \frac{\text{mL}}{0.9982 \text{ g}} = 774 \text{ mL}$$

3.18
$$10.2 \text{ g HCl} \times \frac{\text{mol HCl}}{36.46 \text{ g HCl}} \times \frac{\text{mol } H_2S}{2 \text{ mol HCl}} \times \frac{34.08 \text{ g } H_2S}{\text{mol } H_2S} = 4.77\text{g } H_2S$$

$$13.2 \text{ g FeS} \times \frac{\text{mol FeS}}{87.92 \text{ g FeS}} \times \frac{\text{mol } H_2S}{\text{mol FeS}} \times \frac{34.08 \text{ g } H_2S}{\text{mol } H_2S} = 5.12 \text{ g } H_2S$$

4.77 g H_2S is formed.

3.19
$$20.0 \text{ g alcohol} \times \frac{\text{mol alch}}{88.15 \text{ g alch}} \times \frac{1 \text{ mol acetate}}{1 \text{ mol alch}} \times \frac{130.18 \text{ g acet}}{\text{mol acet}} = 29.5 \text{ g acet}$$

$$25.0 \text{ g acid} \times \frac{\text{mol aa}}{60.05 \text{ g aa}} \times \frac{1 \text{ mol acet}}{1 \text{ mol aa}} \times \frac{130.18 \text{ g acet}}{\text{mol acet}} = 54.20 \text{ g acet}$$

Theoretical yield is 29.5 g isopentyl acetate.

$$\text{Actual Y} = \frac{\%\text{ Y theoret. Y}}{100\%} = \frac{29.5 \text{ g} \times 90.0\%}{100\%} = 26.6 \text{ g}$$

3.20
$$\text{theor Y} = \frac{\text{actual Y} \times 100\%}{\%\text{ Y}} = \frac{433 \text{ g} \times 100\%}{85.0\%} = 509 \text{ g}$$

$$509 \text{ g acet} \times \frac{\text{mol acet}}{130.18 \text{ g acet}} \times \frac{\text{mol alch}}{\text{mol acet}} \times \frac{88.15 \text{ g alch}}{\text{mol alch}} = 345 \text{ g isopentyl alcohol}$$

3.21 a. $\frac{18.0 \text{ mol } H_2SO_4}{2.00 \text{ L solution}} = 9.00 \text{ M } H_2SO_4$

b. $\frac{3.00 \text{ mol KI}}{2.39 \text{ L solution}} = 1.26 \text{ M KI}$

c. $\frac{0.206 \text{ mol HF}}{752 \text{ mL solution}} \times \frac{\text{mL}}{10^{-3} \text{ L}} = 0.274 \text{ M HF}$

d. $\frac{0.522 \text{ g HCl}}{0.592 \text{ L solution}} \times \frac{\text{mol HCl}}{36.46 \text{ g HCl}} = 0.0242 \text{ M HCl}$

e. $\frac{4.98 \text{ g } C_6H_{12}O_6}{224 \text{ mL solution}} \times \frac{\text{mL}}{10^{-3} \text{ L}} \times \frac{\text{mol } C_6H_{12}O_6}{180.16 \text{ g } C_6H_{12}O_6} = 0.123 \text{ M } C_6H_{12}O_6$

f. $\frac{10.5 \text{ g } C_2H_5OH}{24.7 \text{ mL solution}} \times \frac{\text{mL}}{10^{-3} \text{ L}} \times \frac{\text{mol } C_2H_5OH}{46.07 \text{ g } C_2H_5OH} = 9.23 \text{ M } C_2H_5OH$

3.22 a. $2.00 \text{ L} \times 6.00 \text{ M KOH} \times \frac{56.11 \text{ g KOH}}{\text{mol KOH}} = 673 \text{ g KOH}$

b. $100.0 \text{ mL} \times \frac{10^{-3} \text{ L}}{\text{mL}} \times 1.00 \text{ M KOH} \times \frac{56.11 \text{ g KOH}}{\text{mol KOH}} = 5.61 \text{ g KOH}$

c. $10.0 \text{ mL} \times \frac{10^{-3} \text{ L}}{\text{mL}} \times 0.100 \text{ M KOH} \times \frac{56.11 \text{ g KOH}}{\text{mol KOH}} = 0.0561 \text{ g KOH}$

d. $33.0 \text{ mL} \times \frac{10^{-3} \text{ L}}{\text{mL}} \times 2.50 \text{ M KOH} \times \frac{56.11 \text{ g KOH}}{\text{mol KOH}} = 4.63 \text{ g KOH}$

3.23 $\frac{90.0 \text{ g HCOOH}}{100 \text{ g solution}} \times \frac{1.20 \text{ g}}{\text{mL}} \times \frac{\text{mL}}{10^{-3} \text{ L}} \times \frac{\text{mol HCOOH}}{46.03 \text{ g HCOOH}} = 23.5 \text{ M HCOOH}$

3.24 $\frac{15.0 \text{ L} \times 0.315 \text{ M}}{10.15 \text{ M}} \times \frac{\text{mL}}{10^{-3} \text{ L}} = 466 \text{ mL}$

3.25 $15.62 \text{ mL} \times 0.1104 \text{ M } H_2SO_4 \times \frac{2 \text{ mol KOH}}{\text{mol } H_2SO_4} \times \frac{1}{20.00 \text{ mL}} = 0.1724 \text{ M KOH}$

3.26 $750.0 \text{ mL} \times 0.0250 \text{ M } Na_2CrO_4 \times \frac{2 \text{ mol } AgNO_3}{\text{mol } Na_2CrO_4} \times$

$\frac{1}{0.100 \text{ mol } AgNO_3} = 375 \text{ mL } AgNO_3$

Estimation Exercises

3.1 6×10^{23} atoms is 24 g. Thus, $1 \times 10^{23} = 4.0$ g.

$\frac{24.31}{6.022} = 4.04$ actual.

3.2 4 g He is 6.022×10^{23}
Thus, 1 g He is 1.5×10^{23}

3.3 $\frac{40.08 \text{ g}}{\text{mol}} \times \frac{\text{mol}}{6.022 \times 10^{23} \text{ atoms}}$

Estimate $\frac{40}{6} \times 10^{-23} = 6.7 \times 10^{-23}$ g

3.4 SO_4^{2-} $1 \times 32.06 + 4 \times 16.00 = 96.06$ u

NO_3^- $1 \times 14.01 + 3 \times 16.00 = 62.01$ u

I^- = 126.9 u

Two nitrogen atoms per formula unit divided by the smallest formula weight means that NH_4NO_3 has the largest % N.

3.5 The compound with the smallest formula weight, NH_3, will make the most concentrated solution, because there will be more particles in the 10 g mass.

Conceptual Exercise

3.1 $Pb\ (s) + PbO_2\ (s) + 2\ H_2SO_4\ (aq) \rightarrow 2\ PbSO_4\ (s) + 2\ H_2O\ (l)$

Review Questions

3. $1.00 \text{ mol} \times \frac{6.022 \times 10^{23} \text{ molecules}}{\text{mol}} = 6.02 \times 10^{23}$ molecules

$1.00 \text{ mol} \times \frac{6.022 \times 10^{23} \text{ molecules}}{\text{mol}} \times \frac{2 \text{ atoms}}{\text{molecule}} = 1.20 \times 10^{24}$ atoms

4. $1.00 \text{ mol } CaCl_2 \times \frac{6.022 \times 10^{23} \text{ formula units}}{\text{mol}} \times \frac{1\ Ca^{2+}}{\text{formula unit}} = 6.02 \times 10^{23}\ Ca^{2+}$

$1.00 \text{ mol } CaCl_2 \times \frac{6.022 \times 10^{23} \text{ formula units}}{\text{mol}} \times \frac{2\ Cl^-}{\text{formula unit}} = 1.20 \times 10^{24}\ Cl^-$

5. C $1 \times 12.01 = 12.01$
O $2 \times 16.00 = \underline{32.00}$
44.01

The molecular weight is 44.01 u. It is the mass in u of one molecule. The mass of one mole of molecules is the molar mass.

The molar mass is 44.01 g. The atomic weight of C is added to twice the atomic weight of O.

8. a. HO b. CH_2 c. C_5H_4 d. $C_6H_{16}O$

Problems

19. a.
C 6×12.01 u = 72.06 u
H 5×1.008 u = 5.04 u
Br 1×79.90 u = $\underline{79.90 \text{ u}}$
157.00 u

b.
H 3×1.008 u = 3.024 u
P 1×30.9 u = 30.9 u
O 4×15.999 u = $\underline{63.996 \text{ u}}$
97.994 u

c.

K	2 x 39.10 u =	78.20 u
Cr	2 x 52.00 u =	104.00 u
O	7 x 16.00 u =	112.00 u
		294.20 u

d.

Al	2 x 26.98 u =	53.96 u
S	3 x 32.07 u =	96.21 u
O	12 x 16.00 u =	192.00 u
H	36 x 1.008 u =	36.29 u
O	18 x 16.00 u =	288.00 u
		666.46 u

21\. a. $\frac{28.09\ g}{mol} \times \frac{mol}{6.022 \times 10^{23}\ atoms} = 4.665 \times 10^{-23}$ g/atom Si

b. $\frac{63.55\ g}{mol} \times \frac{mol}{6.022 \times 10^{23}\ atoms} = 1.055 \times 10^{-22}$ g/atom Cu

c. $\frac{102.9\ g}{mol} \times \frac{mol}{6.0221 \times 10^{23}\ atoms} = 1.709 \times 10^{-22}$ g/atom Rh

23\. a.

Mn	1 x 54.94 =	54.94
O	2 x 16.00 =	32.00
		86.94

$0.00500\ mol\ MnO_2 \times \frac{86.94\ g}{mol} = 0.435\ g\ MnO_2$

b.

Ca	1 x 40.08 =	40.08
H	2 x 1.008 =	2.02
		42.10

$1.12\ mol\ CaH_2 \times \frac{42.10\ g}{mol} = 47.2\ g\ CaH_2$

c.

C	6 x 12.01 =	72.06
H	12 x 1.008 =	12.10
O	6 x 16.00 =	96.00
		180.16

$0.250\ mol\ C_6H_{12}O_6 \times \frac{180.16\ g}{mol} = 45.0\ g\ C_6H_{12}O_6$

25\. a.

H	1 x 1.008 =	1.01
N	1 x 14.01 =	14.01
O	3 x 16.00 =	48.00
		63.02

$98.6\ g\ HNO_3 \times \frac{mol}{63.02\ g} = 1.56\ mol\ HNO_3$

b.

C	1 x 12.01 =	12.01
Br	4 x 79.90 =	319.60
		331.61

$9.45\ g\ CBr_4 \times \frac{mol}{331.61\ g} = 2.85 \times 10^{-2}\ mol\ CBr_4$

c.

Fe	1 x 55.85 =	55.85
S	1 x 32.06 =	32.07
O	4 x 16.00 =	64.00
		151.92

$$9.11 \text{ g } FeSO_4 \times \frac{\text{mol}}{151.92 \text{ g}} = 6.00 \times 10^{-2} \text{ mol } FeSO_4$$

d.

Pb	1 x 207.2 =	207.2
N	2 x 14.01 =	28.02
O	6 x 16.00 =	96.00
		331.2

$$11.8 \text{ g } Pb(NO_3)_2 \times \frac{\text{mol}}{331.2 \text{ g}} = 3.56 \times 10^{-2} \text{ mol } Pb(NO_3)_2$$

27. a.

Ba	1 x 137.33 =	137.33
Si	1 x 28.09 =	28.09
O	3 x 16.00 =	48.00
		213.42

$$\frac{137.33 \text{ g}}{213.42 \text{ g}} \times 100\% = 64.35\% \text{ Ba}$$

$$\frac{28.09 \text{ g}}{213.42 \text{ g}} \times 100\% = 13.16\% \text{ Si}$$

$$\frac{48.00 \text{ g}}{213.42 \text{ g}} \times 100\% = 22.49\% \text{ O}$$

b.

C	6 x 12.01 =	72.06
H	5 x 1.008 =	5.040
N	1 x 14.01 =	14.01
O	2 x 16.00 =	32.00
		123.11

$$\frac{72.06 \text{ g}}{123.11 \text{ g}} \times 100\% = 58.53\% \text{ C}$$

$$\frac{5.040 \text{ g}}{123.11 \text{ g}} \times 100\% = 4.094\% \text{ H}$$

$$\frac{14.01 \text{ g}}{123.11 \text{ g}} \times 100\% = 11.38\% \text{ N}$$

$$\frac{32.00 \text{ g}}{123.11 \text{ g}} \times 100\% = 25.99\% \text{ O}$$

c.

Mg	1 x 24.31 =	24.31
H	2 x 1.008 =	2.02
C	2 x 12.01 =	24.02
O	6 x 16.00 =	96.00
		146.35

$\frac{24.31\ g}{146.35\ g} \times 100\% = 16.61\%$ Mg

$\frac{2.02\ g}{146.35\ g} \times 100\% = 1.38\%$ H

$\frac{24.02\ g}{146.35\ g} \times 100\% = 16.41\%$ C

$\frac{96.00\ g}{146.35\ g} \times 100\% = 65.60\%$ O

d.

Al	1 x 26.98 =	26.98
Br	3 x 79.90 =	239.70
O	18 x 16.00 =	288.00
H	18 x 1.008 =	18.14
		572.82

$\frac{26.98\ g}{572.82\ g} \times 100\% = 4.71\%$ Al

$\frac{239.70\ g}{572.82\ g} \times 100\% = 41.85\%$ Br

$\frac{288.00\ g}{572.82\ g} \times 100\% = 50.28\%$ O

$\frac{18.14\ g}{572.82\ g} \times 100\% = 3.17\%$ H

29.

C	3 x 12 =	36
H	2 x 1 =	2
Cl	1 x 35 =	35
		73

$\frac{147\ u}{molecule} \times \frac{formula}{73\ u} = \frac{2\ formula}{molecule}$ $C_6H_4Cl_2$ molecular formula

31. $10.05 \text{ g C} \times \frac{\text{mole C}}{12.01 \text{ g C}} = 0.8368 \text{ mol C} \times \frac{1}{0.833 \text{ mol H}} = 1.00$

$0.84 \text{ g H} \times \frac{\text{mole H}}{1.008 \text{ g H}} = 0.833 \text{ mol H} \times \frac{1}{0.833 \text{ mol H}} = 1.0$

$89.10 \text{ g Cl} \times \frac{\text{mole Cl}}{35.45 \text{ g Cl}} = 2.513 \text{ mol Cl} \times \frac{1}{0.833 \text{ mol H}} = 3.02$

$CHCl_3$

33. $65.44 \text{ g C} \times \frac{\text{mol C}}{12.01 \text{ g C}} = 5.449 \text{ mol C} \times \frac{1}{1.816 \text{ mol H}} = 3.000$

$5.49 \text{ g H} \times \frac{\text{mol H}}{1.008 \text{ g H}} = 5.446 \text{ mol H} \times \frac{1}{1.816 \text{ mol H}} = 2.999$

$29.06 \text{ g O} \times \frac{\text{mol O}}{16.00 \text{ g O}} = 1.816 \text{ mol O} \times \frac{1}{1.816 \text{ mol O}} = 1.000$

empirical formula: C_3H_3O

$3 \times 12 + 3 \times 1 + 1 \times 16 = 55$

$\frac{110 \text{ u}}{\text{molecule}} \times \frac{\text{formula}}{55 \text{ u}} = \frac{2 \text{ formula}}{\text{molecule}}$

molecular formula: $C_6H_6O_2$

35. $\frac{0.2147 \text{ g } CO_2}{0.1204 \text{ g sample}} \times \frac{\text{mol } CO_2}{44.01 \text{ g } CO_2} \times \frac{\text{mol C}}{\text{mol } CO_2} \times \frac{12.01 \text{ g C}}{\text{mol C}} \times 100\% = 48.66\% \text{ C}$

$\frac{0.0884 \text{ g } H_2O}{0.1204 \text{ g sample}} \times \frac{\text{mole } H_2O}{18.02 \text{ g } H_2O} \times \frac{2 \text{ mol H}}{\text{mol } H_2O} \times \frac{1.008 \text{ g H}}{\text{mol H}} \times 100\% = 8.214\% \text{ H}$

100.00% - 48.66% C - 8.21% H = 43.13% O

37.
a. $Cl_2O_5 + H_2O \longrightarrow 2\ HClO_3$
b. $V_2O_5 + 2\ H_2 \longrightarrow V_2O_3 + 2\ H_2O$
c. $4\ Al + 3\ O_2 \longrightarrow 2\ Al_2O_3$
d. $2\ C_4H_{10} + 13\ O_2 \longrightarrow 8\ CO_2 + 10\ H_2O$
e. $Sn + 2\ NaOH \longrightarrow Na_2SnO_2 + H_2$
f. $PCl_5 + 4\ H_2O \longrightarrow H_3PO_4 + 5\ HCl$
g. $2\ CH_3OH + 3\ O_2 \longrightarrow 2\ CO_2 + 4\ H_2O$
h. $3\ Zn(OH)_2 + 2\ H_3PO_4 \longrightarrow Zn_3(PO_4)_2 + 6\ H_2O$

39.
a. $2\ Mg(s) + O_2(g) \longrightarrow 2\ MgO(s)$
b. $NH_4NO_3(s) \longrightarrow N_2O(g) + 2\ H_2O(g)$

c. $CH_3CH(OH)CH_2CH_3(l) + 6\ O_2(g) \longrightarrow 4\ CO_2(g) + 5\ H_2O(l)$ (OH drawn above the second carbon)

d. $2\ Al(s) + 6\ HCl(aq) \longrightarrow 2\ AlCl_3(aq) + 3\ H_2(g)$

41. a. $2.0 \times 10^{10}\ mol\ C_8H_{18} \times \frac{16\ mol\ CO_2}{2\ mol\ C_3H_{18}} = 1.6 \times 10^{11}\ mol\ CO_2$

b. $4.4 \times 10^{10}\ mol\ C_8H_{18} \times \frac{25\ mol\ O_2}{2\ mol\ C_8H_{18}} = 5.5 \times 10^{11}\ mol\ O_2$

43. a. $N_2 + 3\ H_2 \longrightarrow 2\ NH_3$

$$440\ g\ H_2 \times \frac{mol\ H_2}{2.016\ g\ H_2} \times \frac{2\ mol\ NH_3}{3\ mol\ H_2} \times \frac{17.03\ g\ NH_3}{mol\ NH_3} = 2.48 \times 10^3\ g\ NH_3$$

b. $N_2 + 3\ H_2 \longrightarrow 2\ NH_3$

$$892\ g\ N_2 \times \frac{mol\ N_2}{28.02\ g\ N_2} \times \frac{3\ mol\ H_2}{mol\ N_2} \times \frac{2.016\ g\ H_2}{mol\ H_2} = 193\ g\ H_2$$

45. a. $C_7H_8 + 3\ HNO_3 \longrightarrow C_7H_5N_3O_6 + 3\ H_2O$

$$454\ g\ C_7H_8 \times \frac{mol\ C_7H_8}{92.13\ g\ C_7H_8} \times \frac{3\ mol\ HNO_3}{mol\ C_7H_8} \times \frac{63.02\ g\ HNO_3}{mol\ HNO_3} = 932\ g\ HNO_3$$

b. $$829\ g\ C_7H_8 \times \frac{mol\ C_7H_8}{92.13\ g\ C_7H_8} \times \frac{mol\ C_7H_5N_3O_6}{mol\ C_7H_8} \times \frac{227.1\ g\ C_7H_5N_3O_6}{mol\ C_7H_5N_3O_6} =$$

$$2.04 \times 10^3\ g\ C_7H_5N_3O_6$$

47. $NH_3 + 2\ O_2 \longrightarrow HNO_3 + H_2O$

$$971\ g\ NH_3 \times \frac{mol\ NH_3}{17.03\ g\ NH_3} \times \frac{mol\ HNO_3}{mol\ NH_3} \times \frac{63.02\ g\ HNO_3}{mol\ HNO_3} = 3.59 \times 10^3\ g\ HNO_3$$

49. a. $2\ C_{14}H_{30} + 43\ O_2 \longrightarrow 28\ CO_2 + 30\ H_2O$

$$1.00\ gal\ C_{14}H_{30} \times \frac{3.785\ L}{1.00\ gal} \times \frac{mL}{10^{-3}\ L} \times \frac{0.763\ g}{mL} \times \frac{mol\ C_{14}H_{30}}{198.38\ g\ C_{14}H_{30}} \times$$

$$\frac{28\ mol\ CO_2}{2\ mol\ C_{14}H_{30}} \times \frac{44.01\ g\ CO_2}{mol\ CO_2} = 8.97 \times 10^3\ g\ CO_2$$

b. $$1.00\ kg\ CO_2 \times \frac{1000\ g}{kg} \times \frac{mol\ CO_2}{44.01\ g\ CO_2} \times \frac{2\ mol\ C_{14}H_{30}}{28\ mol\ CO_2} \times \frac{198.38\ g\ C_{14}H_{30}}{mol\ C_{14}H_{30}} \times$$

$$\frac{mL}{0.763\ g} = 422\ mL\ C_{14}H_{30}$$

51. $$0.150 \text{ mol LiOH} \times \frac{\text{mol } Li_2CO_3}{2 \text{ mol LiOH}} = 0.0750 \text{ mol } Li_2CO_3$$

$$0.080 \text{ mol } CO_2 \times \frac{\text{mol } Li_2CO_3}{\text{mol } CO_2} = 0.080 \text{ mol } Li_2CO_3$$

LiOH is limiting. 0.0750 mol Li_2CO_3 can be produced.

53. $C_3H_8 + 5\ O_2 \longrightarrow 3\ CO_2 + 4\ H_2O$

$$4.81 \text{ g } C_3H_8 \times \frac{\text{mol } C_3H_8}{44.09 \text{ g } C_3H_8} \times \frac{3 \text{ mol } CO_2}{\text{mol } C_3H_8} = 0.3273 \text{ mol } CO_2$$

$$16.4 \text{ g } O_2 \times \frac{\text{mol } O_2}{32.00 \text{ g } O_2} \times \frac{3 \text{ mol } CO_2}{5 \text{ mol } O_2} = 0.3075 \text{ mol } CO_2$$

O_2 is limiting $$0.3075 \text{ mol } CO_2 \times \frac{44.01 \text{ g } CO_2}{\text{mol } CO_2} = 13.5 \text{ g } CO_2$$

55. $$0.488 \text{ g Zn} \times \frac{\text{mol Zn}}{65.39 \text{ g Zn}} \times \frac{8 \text{ mol ZnS}}{8 \text{ mol Zn}} = 0.007463 \text{ mol ZnS}$$

$$0.503 \text{ g } S_8 \times \frac{\text{mol } S_8}{256.56 \text{ g } S_8} \times \frac{8 \text{ mol ZnS}}{\text{mol } S_8} = 0.01568 \text{ mol ZnS}$$

Zn is limiting.

$$0.007463 \text{ mol ZnS} \times \frac{97.46 \text{ g ZnS}}{\text{mol ZnS}} = 0.727 \text{ g ZnS}$$ theoretical yield

$$\frac{0.606 \text{ g ZnS actual}}{0.727 \text{ g ZnS theo}} \times 100\% = 83.4\%$$

57. a. $$14.8 \text{ g } NH_3 \times \frac{\text{mol } NH_3}{17.03 \text{ g } NH_3} \times \frac{\text{mol } NH_4HCO_3}{\text{mol } NH_3} = 0.8691 \text{ mol } NH_4HCO_3$$

$$41.3 \text{ g } CO_2 \times \frac{\text{mol } CO_2}{44.01 \text{ g } CO_2} \times \frac{\text{mol } NH_4HCO_3}{\text{mol } CO_2} = 0.9384 \text{ mol } NH_4HCO_3$$

$$0.8691 \text{ mol } NH_4HCO_3 \times \frac{79.06 \text{ g } NH_4HCO_3}{\text{mol } NH_4HCO_3} \times \frac{74.7 \text{ g}}{100.0\text{g}} = 51.3 \text{ g } NH_4HCO_3$$

b. $$625 \text{ g ZnS} \times \frac{100.0 \text{ g}}{85.0 \text{ g}} \times \frac{\text{mol ZnS}}{97.46 \text{ g ZnS}} \times \frac{\text{mol } Zn(NO_3)_2}{\text{mol ZnS}} \times \frac{189.4 \text{ g } Zn(NO_3)_2}{\text{mol } Zn(NO_3)_2} =$$

$$1.43 \times 10^3 \text{ g } Zn(NO_3)_2$$

59. a. $$\frac{6.00 \text{ mol HCl}}{2.50 \text{ L solution}} = 2.40 \text{ M HCl}$$

b. $$\frac{0.00700 \text{ mol } Li_2CO_3}{10.0 \text{ mL solution}} \times \frac{\text{mL}}{10^{-3} \text{ L}} = 0.700 \text{ M } Li_2CO_3$$

61. a. $\frac{8.90 \text{ g } H_2SO_4}{100.0 \text{ mL solution}} \times \frac{\text{mol } H_2SO_4}{98.09 \text{ g } H_2SO_4} \times \frac{\text{mL}}{10^{-3} \text{ L}} = 0.907 \text{ M } H_2SO_4$

b. $\frac{439 \text{ g } C_6H_{12}O_6}{1.25 \text{ L solution}} \times \frac{\text{mol } C_6H_{12}O_6}{180.2 \text{ g } C_6H_{12}O_6} = 1.95 \text{ M } C_6H_{12}O_6$

63. a. $2.00 \text{ L} \times 1.00 \text{ M} \times \frac{40.00 \text{ g NaOH}}{\text{mol NaOH}} = 80.0 \text{ g NaOH}$

b. $10.0 \text{ mL} \times \frac{10^{-3} \text{ L}}{\text{mL}} \times 4.25 \text{ M} \times \frac{180.2 \text{ g } C_6H_{12}O_6}{\text{mol } C_6H_{12}O_6} = 7.66 \text{ g } C_6H_{12}O_6$

65. $1.25 \text{ mol NaOH} \times \frac{1}{6.00 \text{ M}} \times \frac{\text{mL}}{10^{-3} \text{ L}} = 208 \text{ mL}$

67. $8.10 \text{ g } KMnO_4 \times \frac{\text{mol } KMnO_4}{158.0 \text{ g } KMnO_4} \times \frac{1}{0.0250 \text{ M}} \times \frac{\text{mL}}{10^{-3} \text{ L}} = 2.05 \times 10^3 \text{ mL}$

69. $2.00 \text{ L} \times \frac{1.00 \text{ mol}}{\text{L}} \times \frac{1 \text{ L}}{12.0 \text{ mol}} \times \frac{\text{mL}}{10^{-3} \text{ L}} = 167 \text{ mL}$

71. $25.00 \text{ mL} \times \frac{10^{-3} \text{ L}}{\text{mL}} \times \frac{1.04 \text{ mol}}{\text{L}} \times \frac{1}{0.500 \text{ L}} = 0.0520 \text{ M}$

73. (c) 0.17 M NH_3: The concentration must be greater than the average of 0.10 M and 0.02 M, but less than their sum (0.30 M).
The actual calculation follows:

$$C = \frac{(0.100 \text{ L} \times 0.100 \text{ mol/L}) + (0.200 \text{ L} \times 0.200 \text{ mol/L})}{0.100 \text{ L} + 0.200 \text{ L}} = 0.167 \text{ M} \Rightarrow 0.17 \text{ M}$$

75. $\frac{67.0 \text{ g } HNO_3}{100 \text{ g solution}} \times \frac{1.40 \text{ g}}{\text{mL}} \times \frac{\text{mL}}{10^{-3} \text{ L}} \times \frac{\text{mol } HNO_3}{63.02 \text{ g } HNO_3} = 14.9 \text{ M } HNO_3$

77. $33.22 \text{ mL} \times \frac{10^{-3} \text{ L}}{\text{mL}} \times 0.1503 \text{ M NaOH} \times \frac{\text{mol HCl}}{\text{mol NaOH}} \times \frac{1}{20.00 \text{ mL} \times \frac{10^{-3} \text{ L}}{\text{mL}}}$

$= 0.2496 \text{ M HCl}$

79. $28.27 \text{ mL} \times \frac{10^{-3} \text{ L}}{\text{mL}} \times 0.01025 \text{ M HCl} \times \frac{\text{mol } Ca(OH)_2}{2 \text{ mol HCl}} \times \frac{1}{18.50 \text{ mL} \times \frac{10^{-3} \text{ L}}{\text{mL}}}$

$= 7.832 \times 10^{-3} \text{ M } Ca(OH)_2$

81. $10.32 \text{ mL} \times \frac{10^{-3} \text{ L}}{\text{mL}} \times 0.4042 \text{ M } NaHCO_3 \times \frac{\text{mol } H_2SO_4}{2 \text{ mol } NaHCO_3} \times \frac{1}{0.1000 \text{ M } H_2SO_4}$

$\times \frac{\text{mL}}{10^{-3} \text{ L}} = 20.86 \text{ mL}$

83. It is necessary to make 100 mL of solution, as that is the smallest volumetric flask available with a volume greater than 80 mL. Measure 100.0 mL of the 0.04000 M $AgNO_2$ solution in the 100.0 mL volumetric flask and add 1.0194g of $AgNO_3$.

$$100.0 \text{ mL} \times \frac{10^{-3}\text{ L}}{\text{mL}} \times 1.000\text{ M} - 100.0\text{ mL} \times \frac{10^{-3}\text{ L}}{\text{mL}} \times 0.04000\text{ M}$$
$$= 6.000 \times 10^{-3}\text{ mol needed}$$

$$6.000 \times 10^{-3}\text{ mol} \times \frac{10^{-3}\text{ mol}}{\text{mmol}} \times \frac{169.88\text{ g AgNO}_3}{\text{mol AgNO}_3} = 1.0194\text{ g AgNO}_3\text{ added.}$$

85. a. $$875\text{ mg Ca} \times \frac{100.1\text{ g CaCO}_3}{40.08\text{ g Ca}} \times \frac{10^{-3}\text{ g}}{\text{mg}} = 2.19\text{ g CaCO}_3$$

b. $$875\text{ mg Ca} \times \frac{218.2\text{ g Ca(C}_3\text{H}_5\text{O}_3)_2}{40.08\text{ g Ca}} \times \frac{10^{-3}\text{ g}}{\text{mg}} = 4.76\text{ g Ca(C}_3\text{H}_5\text{O}_3)_2$$

c. $$875\text{ mg Ca} \times \frac{430.4\text{ g Ca(C}_6\text{H}_{11}\text{O}_7)_2}{40.08\text{ g Ca}} \times \frac{10^{-3}\text{ g}}{\text{mg}} = 9.40\text{ g Ca(C}_6\text{H}_{11}\text{O}_7)_2$$

d. $$875\text{ mg Ca} \times \frac{498.4\text{ g Ca}_3\text{(C}_6\text{H}_5\text{O}_7)_2}{3 \times 40.08\text{ g Ca}} \times \frac{10^{-3}\text{ g}}{\text{mg}} = 3.63\text{ g Ca}_3\text{(C}_6\text{H}_5\text{O}_7)_2$$

87. $$\frac{24.31\text{ u Mg}}{\text{atom Mg}} \times \frac{\text{atom Mg}}{\text{molecule chlorophyll}} \times \frac{100\text{ u chlorophyll}}{2.72\text{ u Mg}} = \frac{894\text{ u}}{\text{molecule}}$$

89. a. $$\frac{0.6260\text{ g CO}_2}{0.1888\text{ g sample}} \times \frac{12.011\text{ g C}}{44.010\text{ g CO}_2} \times 100\% = 90.49\%\text{ C}$$

$$\frac{0.1602\text{ g H}_2\text{O}}{0.1888\text{ g sample}} \times \frac{2.0159\text{ g H}}{18.015\text{ g H}_2\text{O}} \times 100\% = 9.495\%\text{ H}$$

b. $$90.48\text{ g C} \times \frac{\text{mol C}}{12.011\text{ g C}} = 7.533\text{ mol C} \times \frac{1}{7.533\text{ mol C}} = 1.000 \times 4 = 4.00$$

$$9.495\text{ g H} \times \frac{\text{mol H}}{1.0079\text{ g H}} = 9.481\text{ mol H} \times \frac{1}{7.533\text{ mol C}} = 1.251 \times 4 = 5.00$$

C_4H_5 empirical formula

c. $$\frac{106\text{ u}}{\text{molecule}} \times \frac{\text{formula}}{53.08\text{ u}} = \frac{2\text{ formula}}{\text{molecule}}$$

C_8H_{10} molecular formula

91. $3.047 \text{ g } CO_2 \times \frac{12.011 \text{ g C}}{44.009 \text{ g } CO_2} = 0.8316 \text{ g C}$

$1.247 \text{ g } H_2O \times \frac{2.0158 \text{ g H}}{18.015 \text{ g } H_2O} = 0.1395 \text{ g H}$

1.525 g sample - 0.832 g C - 0.140 g H = 0.553 g O

$0.553 \text{ g O} \times \frac{\text{mol O}}{16.00 \text{ g O}} = 3.456 \times 10^{-2} \text{ mol O} \times \frac{1}{0.03456 \text{ mol O}} = 1.00$

$0.832 \text{ g C} \times \frac{\text{mol C}}{12.01 \text{ g C}} = 6.928 \times 10^{-2} \text{ mol C} \times \frac{1}{0.03456 \text{ mol O}} = 2.00$

$0.1395 \text{ g H} \times \frac{\text{mol H}}{1.008 \text{ g H}} = 1.384 \times 10^{-1} \text{ mol H} \times \frac{1}{0.03456 \text{ mol O}} = 4.00$

C_2H_4O formula weight 44.0 u

$\frac{88.1 \text{ u}}{\text{molecule}} \times \frac{\text{formula}}{44.0 \text{ u}} = \frac{2 \text{ formula}}{\text{molecule}}$

$C_4H_8O_2$

There are several structures that could be made of $C_4H_8O_2$. Some of them are:

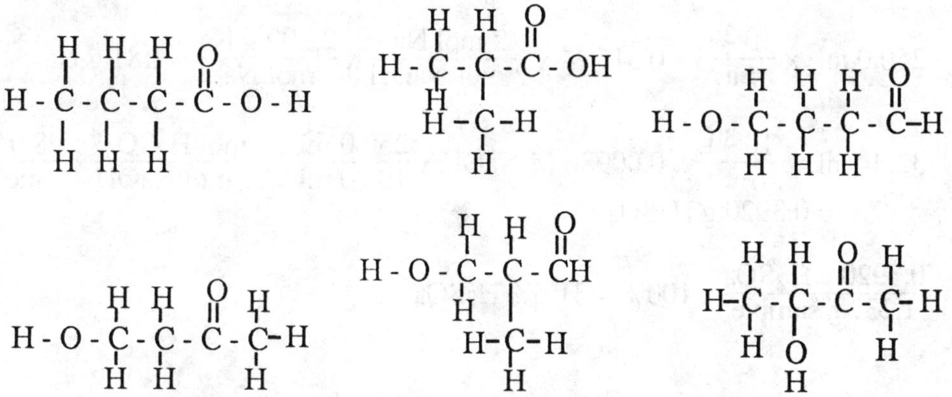

93. $$2\,Al + 6\,HCl \longrightarrow 2\,AlCl_3 + 3\,H_2$$

$12.3 \text{ cm} \times 14.3 \text{ cm} \times 2.2 \text{ mm} \times \frac{10^{-3} \text{ m}}{\text{mm}} \times \frac{\text{cm}}{10^{-2} \text{ m}} \times \frac{2.70 \text{ g}}{\text{cm}^3} \times \frac{\text{mol Al}}{26.98 \text{ g Al}} \times \frac{3 \text{ mol } H_2}{2 \text{ mol Al}}$

$\times \frac{2.016 \text{ g } H_2}{\text{mol } H_2} = 12 \text{ g } H_2$

95. $228 \text{ trains} \times \frac{115 \text{ cars}}{\text{train}} \times \frac{90.5 \text{ mt}}{\text{car}} \times \frac{10^3 \text{ kg}}{\text{mt}} \times \frac{10^3 \text{ g}}{\text{kg}} \times \frac{64.3 \text{ g C}}{100 \text{ g coal}} \times$
$\frac{\text{mol C}}{12.01 \text{ g C}} \times \frac{\text{mol } CO_2}{\text{mol C}} \times \frac{44.01 \text{ g } CO_2}{\text{mol } CO_2} \times \frac{1 \text{ kg } CO_2}{10^3 \text{ g } CO_2} \times \frac{\text{mt } CO_2}{1000 \text{ kg}}$
$= 5.59 \times 10^6$ metric tons CO_2

97. $13.0 \text{ g} \times \frac{\text{mol } C_4H_9OH}{74.12 \text{ g } C_4H_9OH} \times \frac{\text{mol } C_4H_9Br}{\text{mol } C_4H_9OH} = 0.1754 \text{ mol } C_4H_9Br$

$21.6 \text{ g} \times \frac{\text{mol NaBr}}{102.9 \text{ g NaBr}} \times \frac{\text{mol } C_4H_9Br}{\text{mol NaBr}} = 0.2099 \text{ mol } C_4H_9Br$

$33.8 \text{ g} \times \frac{\text{mol } H_2SO_4}{98.09 \text{ g } H_2SO_4} \times \frac{\text{mol } C_4H_9Br}{\text{mol } H_2SO_4} = 0.3446 \text{ mol } C_4H_9Br$

C_4H_9OH limiting $0.1754 \text{ mol } C_4H_9Br \times \frac{137.0 \text{ g } C_4H_9Br}{\text{mol } C_4H_9Br} = 24.0$ g theoretical yield

16.8 g actual yield

$\frac{16.8 \text{ g}}{24.0 \text{ g}} \times 100\% = 70.0\%$ yield

99. $250.0 \text{ mL} \times \frac{10^{-3} \text{ L}}{\text{mL}} \times 0.315 \text{ M} \times \frac{2 \text{ mol Na}}{2 \text{ mol NaOH}} \times \frac{22.99 \text{ g Na}}{\text{mol Na}} = 1.81 \text{ g Na}$

101. $32.44 \text{ mL} \times \frac{10^{-3} \text{ L}}{\text{mL}} \times 0.00986 \text{ M NaOH} \times \frac{250.0 \text{ mL}}{10.00 \text{ mL}} \times \frac{\text{mol } H_2SO_4}{2 \text{ mol NaOH}} \times \frac{98.04 \text{ g}}{\text{mol}}$
$= 0.3920 \text{ g } H_2SO_4$

$\frac{0.3920 \text{ g } H_2SO_4}{1.239 \text{ g sample}} \times 100\% = 31.6\% \ H_2SO_4$

Chapter 4

Gases

Exercises

4.1 a. $722\text{ torr} \times \frac{1\text{ mmHg}}{\text{torr}} = 722\text{ mmHg}$

b. $98.2\text{ kPa} \times \frac{760\text{ torr}}{101.325\text{ kPa}} = 737\text{ torr}$

c. $29.95\text{ in.Hg} \times \frac{2.54\text{ cm}}{1.000\text{ in}} \times \frac{10^{-2}\text{ m}}{\text{cm}} \times \frac{\text{mm}}{10^{-3}\text{ m}} \times \frac{\text{torr}}{\text{mmHg}} = 760.7\text{ torr}$

d. $768\text{ torr} \times \frac{\text{atm}}{760\text{ torr}} = 1.01\text{ atm}$

4.2 $h_{CCl_4} d_{CCl_4} = d_{Hg}\, h_{Hg}$

$$h_{CCl_4} = \frac{d_{Hg}\, h_{Hg}}{d_{CCl_4}} = \frac{\frac{13.6\text{ g}}{\text{cm}^3}}{\frac{1.59\text{ g}}{\text{cm}^3}} \times 760\text{ mm} = 6.50 \times 10^3\text{ mm} = 6.50\text{ m}$$

4.3 $$h_{Hg} = \frac{d_{H_2O} \times h_{H_2O}}{d_{Hg}} = 30.0\text{ m} \times \frac{\text{mm}}{10^{-3}\text{ m}} \times \frac{1.00\text{ g/cm}^3}{13.6\text{ g/cm}^3} = 2.21 \times 10^3\text{ mmHg}$$

$$P_{H_2O} = 2.21 \times 10^3\text{ mmHg} \times \frac{\text{atm}}{760\text{ mmHg}} = 2.90\text{ atm}$$

4.4 $P_1V_1 = P_2V_2$ $\quad V_2 = \frac{P_1V_1}{P_2} = \frac{98.7\text{ kPa} \times 73.3\text{ mL}}{4.02\text{ atm}} \times \frac{\text{atm}}{101.325\text{ kPa}} = 17.8\text{ mL}$

4.5 $P_1V_1 = P_2V_2$ $\quad P_2 = \frac{P_1V_1}{V_2} = \frac{535\text{ mL} \times 988\text{ torr}}{1.05\text{ L} \times \frac{\text{mL}}{10^{-3}\text{ L}}} = 503\text{ torr}$

4.6 $\frac{V_1}{T_1} = \frac{V_2}{T_2}$ $\quad V_2 = \frac{V_1T_2}{T_1} = \frac{692\text{ L} \times (273 + 23)\text{K}}{(273 + 602)\text{K}} = 234\text{ L}$

4.7 $5.00 \times 10^3\text{ L} \times \frac{\text{mol}}{22.4\text{ L}} \times \frac{44.01\text{ g}}{\text{mol}} \times \frac{\text{kg}}{10^3\text{ g}} = 9.82\text{ kg}$

4.8 $\frac{P_1V_1}{T_1} = \frac{P_2V_2}{T_2}$ $\quad V_1 = V_2$

$$T_2 = \frac{P_2T_1}{P_1} = \frac{8.0\text{ atm} \times (273 + 22)\text{K}}{2.5\text{ atm}}$$

$$T_2 = 944\text{ K} - 273 = 671\text{ °C}$$

4.9 $\dfrac{P_1V_1}{n_1T_1} = \dfrac{P_2V_2}{n_2T_2}$, $T_1 = T_2$ and $V_1 = V_2$

$\dfrac{P_1}{n_1} = \dfrac{P_2}{n_2}$ or $\dfrac{P}{n}$ = constant

If the number of molecules increases, the pressure must increase as more molecular collisions with the wall occur.

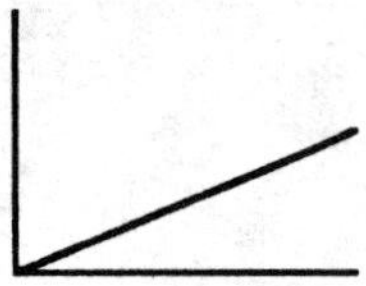

4.10 $n = \dfrac{PV}{RT} = \dfrac{3.15 \text{ atm} \times 35.0 \text{ L}}{\dfrac{0.08206 \text{ L atm}}{\text{K mol}} \times 852 \text{ K}} = 1.58 \text{ moles}$

4.11 $T = \dfrac{PV\mathcal{M}}{mR} = \dfrac{785 \text{ torr} \times \dfrac{\text{atm}}{760 \text{ torr}} \times 5.00 \text{ L} \times 32.00 \text{ g}}{15.0 \text{ g} \times \dfrac{0.08206 \text{ L atm}}{\text{K mol}}} = 134 \text{ K}$

134 K -273 = -139 °C where $\mathcal{M}$ is the molar mass and $n = \dfrac{m}{\mathcal{M}}$

4.12 $m = \dfrac{PV\mathcal{M}}{RT} = \dfrac{546 \text{ torr} \times 525 \text{ mL} \times \dfrac{28.02 \text{ g}}{\text{mol}} \times \dfrac{10^{-3} \text{ L}}{\text{mL}} \times \dfrac{\text{atm}}{760 \text{ torr}}}{\dfrac{0.08206 \text{ L atm}}{\text{K mol}} \times (35 + 273) \text{ K}}$ $m = 0.418$ g

4.13 $\mathcal{M} = \dfrac{mRT}{PV} = \dfrac{0.440 \text{ g} \times \dfrac{0.08206 \text{ L atm}}{\text{K mol}} \times (86 + 273) \text{ K}}{741 \text{ mmHg} \times 179 \text{ mL} \times \dfrac{10^{-3} \text{ L}}{\text{mL}} \times \dfrac{\text{atm}}{760 \text{ mmHg}}}$ M = 74.3 g/mol

4.14 $\mathcal{M} = \dfrac{mRT}{PV} = \dfrac{0.471 \text{ g} \times \dfrac{0.08206 \text{ L atm}}{\text{K mol}} \times (98 + 273) \text{ K}}{715 \text{ mmHg} \times 121 \text{ mL} \times \dfrac{10^{-3} \text{ L}}{\text{mL}} \times \dfrac{\text{atm}}{760 \text{ mmHg}}}$ M = 126 g/mol

4.15 $55.80 \text{ g C} \times \dfrac{\text{mol C}}{12.011 \text{ g C}} = 4.646 \text{ mol C} \times \dfrac{1}{2.323 \text{ mol}} = 2.00$

$7.03 \text{ g H} \times \dfrac{\text{mol H}}{1.008 \text{ g H}} = 6.974 \text{ mol H} \times \dfrac{1}{2.323 \text{ mol}} = 3.00$

$37.17 \text{ g O} \times \dfrac{\text{mol O}}{16.00 \text{ g O}} = 2.323 \text{ mol O} \times \dfrac{1}{2.323 \text{ mol O}} = 1.00$

Empirical formula is C_2H_3O, at 43.04 g/formula unit.

$$\mathcal{M} = \frac{mRT}{PV} = \frac{0.3060\text{ g} \times \frac{0.08206\text{ L atm}}{\text{K mol}} \times (100 + 273)\text{ K}}{747\text{ mmHg} \times 111\text{ mL} \times \frac{10^{-3}\text{ L}}{\text{mL}} \times \frac{\text{atm}}{760\text{ mmHg}}} \qquad \mathcal{M} = 85.8\text{ g/mol}$$

$$\frac{85.8\text{ g}}{\text{mol}} \times \frac{\text{formula}}{43.03\text{ g}} = 2\text{ formula/mol}$$

Molecular formula is $C_4H_6O_2$.

4.16 $PV = nRT$

$$P = \frac{mRT}{V\mathcal{M}} = \frac{dRT}{\mathcal{M}}$$

$$d = \frac{P\mathcal{M}}{RT} = \frac{748\text{ torr} \times 30.07\text{ g/mol} \times \frac{\text{atm}}{760\text{ torr}}}{\frac{0.08206\text{ L atm}}{\text{K mol}} \times (15 + 273)\text{ K}} = 1.25\text{ g/L}$$

4.17 $$T = \frac{P\mathcal{M}}{Rd} = \frac{785\text{ torr} \times 44.09\text{ g/mol} \times \frac{\text{atm}}{760\text{ torr}}}{\frac{0.08206\text{ L atm}}{\text{K mol}} \times 1.51\text{ g/L}} = 367\text{ K}$$

367 K - 273 = 94 °C

4.18 $$10.0\text{ L }H_2 \times \frac{\text{L }CH_4}{2\text{ L }H_2} = 5.00\text{ L }CH_4$$

4.19 $$V = \frac{nRT}{P}$$

$$45.8\text{ kg }CaCO_3 \times \frac{10^3\text{ g}}{\text{kg}} \times \frac{1\text{ mol }CaCO_3}{100.09\text{ g }CaCO_3} \times \frac{\text{mol }CO_2}{\text{mol }CaCO_3} \times \frac{\frac{0.08206\text{ L atm}}{\text{K mol}} \times (825 + 273)\text{ K}}{754\text{ torr} \times \frac{\text{atm}}{760\text{ torr}}} = V = 4.16 \times 10^4\text{ L}$$

4.20 $$\frac{PV}{RT} = n_{CO_2}$$

$$\frac{733\text{ torr} \times 1.25 \times 10^4\text{ L} \times \frac{\text{atm}}{760\text{ torr}}}{\frac{0.08206\text{ L atm}}{\text{K mol}} \times (825 + 273)\text{ K}} \times \frac{\text{mol CaO}}{\text{mol }CO_2} \times \frac{56.08\text{ g CaO}}{\text{mol CaO}} \times \frac{\text{kg}}{10^3\text{ g}}$$

= 7.50 kg CaO

4.21 $P_{O_2} = \frac{n_{O_2}RT}{V} = \frac{0.00856 \text{ mol x } \frac{0.08206 \text{ L atm}}{\text{mol K}} \text{ x } 298 \text{ K}}{1.00 \text{ L}}$

$P_{O_2} = 0.209 \text{ atm}$

$P_{Ar} = \frac{n_{Ar}RT}{V} = \frac{0.000381 \text{ mol x } 0.08206 \text{ L mol K atm x } 298 \text{ K}}{1.00 \text{ L}} = 0.00932 \text{ atm}$

$P_{CO_2} = \frac{n_{CO_2}RT}{V} = \frac{0.00002 \text{ mol x } \frac{0.08206 \text{ L atm}}{\text{mol K}} \text{ x } 298 \text{ K}}{1.00 \text{ L}} = 0.0005 \text{ atm}$

$P_{total} = P_{N_2} + P_{O_2} + P_{Ar} + P_{CO_2} = (0.780 + 0.209 + 0.00932 + 0.0005) \text{ atm}$

$P_{total} = 0.999 \text{ atm}$

4.22 $P_{N_2} = \frac{m_{N_2}RT}{\mathcal{M}V} = \frac{4.05 \text{ g x } \frac{0.08206 \text{ L atm}}{\text{K mol}} \text{ x } (25 + 273) \text{ K}}{28.01 \text{ g/mol x } 6.10 \text{ L}}$

$P_{N_2} = 0.580 \text{ atm}$

$P_{H_2} = \frac{m_{H_2}RT}{\mathcal{M}V} = \frac{3.15 \text{ g x } \frac{0.08206 \text{ L atm}}{\text{K mol}} \text{ x } (25 + 273) \text{ K}}{2.016 \text{ g/mol x } 6.10 \text{ L}}$

$P_{H_2} = 6.26 \text{ atm}$

$P_{He} = \frac{m_{He}RT}{\mathcal{M}V} = \frac{6.05 \text{ g x } \frac{0.08206 \text{ L atm}}{\text{K mol}} \text{ x } (25 + 273) \text{ K}}{4.003 \text{ g/mol x } 6.10 \text{ L}}$

$P_{He} = 6.06 \text{ atm}$

$P_{total} = P_{N_2} + P_{H_2} + P_{He} = 12.90 \text{ atm}$

4.23 $P_{N_2} = 0.741 \text{ x } 1.000 \text{ atm} = 0.741 \text{ atm}$

$P_{O_2} = 0.150 \text{ x } 1.000 \text{ atm} = 0.150 \text{ atm}$

$P_{H_2O} = 0.060 \text{ x } 1.000 \text{ atm} = 0.060 \text{ atm}$

$P_{Ar} = 0.009 \text{ x } 1.000 \text{ atm} = 0.009 \text{ atm}$

$P_{CO_2} = 0.040 \text{ x } 1.000 \text{ atm} = 0.040 \text{ atm}$

4.24 $P_{total} = P_{CH_4} + P_{C_2H_6} + P_{C_3H_8} + P_{C_4H_{10}}$

$P_{total} = (505 + 201 + 43 + 11.2)\text{ torr} = 760\text{ torr}$

$$\chi_{CH_4} = \frac{505\text{ torr}}{760\text{ torr}} = 0.664$$
$$\chi_{C_2H_6} = \frac{201\text{ torr}}{760\text{ torr}} = 0.264$$
$$\chi_{C_3H_8} = \frac{43\text{ torr}}{760\text{ torr}} = 0.057$$
$$\chi_{C_4H_{10}} = \frac{11.2\text{ torr}}{760\text{ torr}} = 0.015$$

4.25 $P_{total} = P_{N_2} + P_{H_2O}$

$P_{N_2} = 696\text{ torr} - 19\text{ torr} = 677\text{ torr}$

$$V = \frac{mRT}{\mathcal{M}P} = \frac{1.28\text{ g} \times \frac{0.08206\text{ L atm}}{\text{K mol}} \times (21 + 273)\text{ K}}{\frac{28.01\text{ g}}{\text{mol}} \times 677\text{ torr} \times \frac{\text{atm}}{760\text{ torr}}}$$

$V = 1.24\text{ L}$

4.26 $P_{H_2} = P_{total} - P_{H_2O} = 738\text{ torr} - 16\text{ torr} = 722\text{ torr}$

$$m_{H_2} = \frac{\mathcal{M}PV}{RT} = \frac{2.016\text{ g/mol} \times 722\text{ torr} \times 246\text{ mL} \times \frac{10^{-3}\text{ L}}{\text{mL}} \times \frac{\text{atm}}{760\text{ torr}}}{\frac{0.08206\text{ L atm}}{\text{K mol}} \times (18 + 273)\text{ K}} = 0.0197\text{ g }H_2$$

$$m_{H_2O} = \frac{18.02\text{ g/mol} \times 15.5\text{ torr} \times 246\text{ mL} \times \frac{10^{-3}\text{ L}}{\text{mL}} \times \frac{\text{atm}}{760\text{ torr}}}{\frac{0.08206\text{ L atm}}{\text{K mol}} \times (18 + 273)\text{K}} = 0.00379\text{ g }H_2O$$

$m_{total} = 0.0235\text{ g}$

4.27 $2\ KClO_3 \rightarrow 2\ KCl + 3\ O_2$

$P_{H_2} = P_{total} - P_{H_2O} = 746\text{ mmHg} - 19\text{ mmHg} = 727\text{ mmHg}$

$$\frac{727\text{ mmHg} \times 155\text{ mL} \times \frac{10^{-3}\text{ L}}{\text{mL}} \times \frac{\text{atm}}{760\text{ mmHg}}}{\frac{0.08206\text{ L atm}}{\text{K mol}} \times (21 + 273)\text{ K}} \times \frac{2\text{ mol }KClO_3}{3\text{ mol }O_2} \times \frac{122.55\text{ g}}{\text{mol}}$$

$= 0.502\text{ g }KClO_3$

4.28 $\frac{r_{N_2}}{r_{Ar}} = \sqrt{\frac{M_{Ar}}{M_{N_2}}} = \sqrt{\frac{39.95}{28.01}} = 1.19$

4.29 $\frac{r_{N_2}}{r_{unk}} = \frac{t_{unk}}{t_{N_2}} = \frac{83\text{ s}}{57\text{ s}} = \sqrt{\frac{\mathcal{M}_{unk}}{\mathcal{M}_{N_2}}} = 1.46$

$\mathcal{M}_{unk} = (1.46)^2 \times 28.01\text{ g/mol} = 60\text{ g/mol}$

4.30 $\frac{t_{CH_4}}{t_{O_2}} = \sqrt{\frac{M_{CH_4}}{M_{O_2}}} \quad \frac{t_{CH_4}}{123\text{ s}} = \sqrt{\frac{16.04}{32.00}}$

$t_{CH_4} = 87.1\text{ s}$

Estimation Exercises

4.1 $P_1V_1 = P_2V_2$

$$P_2 = \frac{P_1V_1}{V_2} = \frac{10.2\text{ L} \times 1208\text{ torr}}{30.0\text{ L}}$$

10.2 L is about $\frac{1}{3}$ of 30, so $\frac{1}{3}$ of 1208 torr is about 400 torr or 400 mmHg.

4.2 $\frac{V_1}{T_1} = \frac{V_2}{T_2} \qquad \frac{T_2}{T_1} = \frac{V_2}{V_1}$

To double the volume from 2.50 L to 5.00 L, the Kelvin temperature must be doubled.

$$\text{estimate} = \frac{5.00\text{ L}}{2.50\text{ L}} \times 150\text{ K} = 300\text{ K}$$

$T = 300\text{ K} - 273 \sim 30\text{ °C}$ actual 27 °C

4.3 At STP, 1 mole is 22.4 L. $H_2 = \frac{2\text{ g}}{\text{mol}}$; He = 4 g/mol; CH_4 = 16 g/mol.

H_2 is hotter and at less pressure, so it will be less dense than at STP. $d = \frac{P\mathcal{M}}{RT}$.
CH_4 is colder and at higher pressure, so it is more dense than at STP.
CH_4 has the greatest density.

4.4 The O_2-to-H_2 molar mass ratio is 32/2 = 16. For u_{rms} of O_2 to be greater than 1838 m/s (u_{rms} of H_2 at 0 °C), $T/16$ must be greater than 273 K.

T must be greater than 16 x 273, which simplifies to 15 x 300 = 4500 K. Answer: 5000 K; actual: 16 x 273 = 4368 K.

Conceptual Exercises

4.1 The increase to 3.00 atm could be achieved by adding only hydrogen, but the P_{H_2} would be 2.50 atm, not 2.00 atm. To achieve a 3.00 atm pressure with P_{H_2} = 2.00 atm, other gases must supply 1.00 atm pressure. The original He supplies only 0.50 atm.

4.2

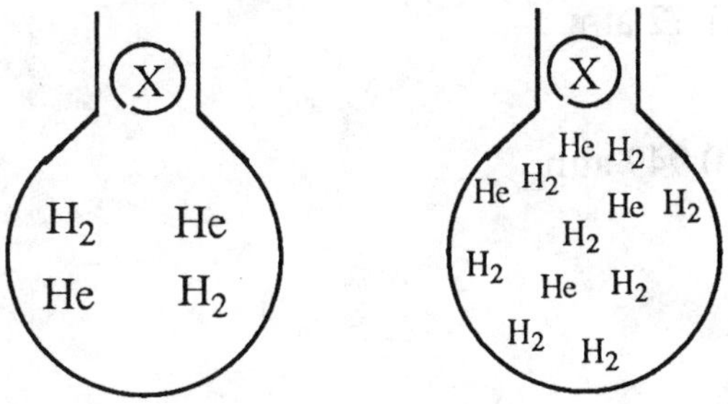

Review Questions

13. $T_K = T_C + 273$

14. a. The volume will decrease.
b. The volume will decrease.
c. The volume will increase.

15. a. The pressure will increase.
b. The pressure will increase.
c. The pressure will increase.

16. a. The temperature is decreasing.
b. The pressure decreases.

18. a. A; A higher pressure in container A means more molecules and thus more mass and a higher density.
b. They are equal. Equal pressures and temperatures mean the same number of molecules per volume.
c. B; A higher temperature means that container A has fewer molecules and thus less mass and a lower density.

19. T = 273 K P = 1 atm
A given amount of any gas will be the same volume at equal pressure and temperature. If everyone picks the same standard conditions, it is much easier to compare experiments.

25. $$\chi_A = \frac{n_A}{n_A + n_B + \ldots\ldots} = \frac{n_A}{n_{total}}$$

$$\chi_A = \frac{P_A}{P_A + P_B + \ldots\ldots} = \frac{P_A}{P_{total}}$$

Problems

31. $$4.36 \text{ atm x } \frac{760 \text{ mmHg}}{\text{atm}} = 3.31 \text{ x } 10^3 \text{ mmHg}$$

33. a. $$0.985 \text{ atm x } \frac{760 \text{ mmHg}}{\text{atm}} = 749 \text{ mmHg}$$

b. $$849 \text{ torr x } \frac{\text{atm}}{760 \text{ torr}} = 1.12 \text{ atm}$$

c. $$721 \text{ torr x } \frac{1 \text{ atm}}{760 \text{ torr}} = 0.949 \text{ atm}$$

35. a. $$h_{oil} \text{ x } \frac{d_{oil}}{d_{Hg}} = h_{Hg}$$

$$44 \text{ mm oil x } \frac{0.798 \text{ g/mL}}{13.6 \text{ g/mL}} = 2.58 \text{ mmHg}$$

755 mmHg + 3 mmHg = 758 mmHg

b. $$h \text{ x } \frac{d_{oil}}{d_{Hg}} = h_{Hg}$$

$$22.3 \text{ cm x } \frac{10 \text{ mm}}{\text{cm}} \text{ x } \frac{0.798 \text{ g/mL}}{13.6 \text{ g/mL}} = 13.1 \text{ mmHg}$$

735 mmHg - 13 mmHg = 722 mmHg

37. $P_1V_1 = P_2V_2$

a. $$V_2 = \frac{P_1V_1}{P_2} = 521 \text{ mL x } \frac{1572 \text{ torr}}{752 \text{ torr}} = 1.09 \text{ x } 10^3 \text{ mL x } \frac{10^{-3} \text{ L}}{\text{mL}} = 1.09 \text{ L}$$

b. $$V_2 = \frac{P_1V_1}{P_2} = 521 \text{ mL x } \frac{1572 \text{ torr}}{3.55 \text{ atm x } \frac{760 \text{ torr}}{\text{atm}}} = 304 \text{ mL}$$

c. $$P_2 = \frac{P_1V_1}{V_2} = \frac{521 \text{ mL}}{315 \text{ mL}} \text{ x } 1572 \text{ torr} = 2.60 \text{ x } 10^3 \text{ torr}$$

d. $$P_2 = \frac{P_1V_1}{V_2} = \frac{521 \text{ mL}}{2.75 \text{ L}} \text{ x } \frac{10^{-3} \text{ L}}{\text{mL}} \text{ x } 1572 \text{ torr x } \frac{\text{atm}}{760 \text{ torr}} = 0.392 \text{ atm}$$

39. $P_1V_1 = P_2V_2$ $$V_2 = \frac{P_1V_1}{P_2} = \frac{1070 \text{ psi}}{14.7 \text{ psi}} \text{ x } 19 \text{ x } 10^6 \text{ ft}^3 = 1.4 \text{ x } 10^9 \text{ ft}^3$$

41. $A = \pi r^2 = 3.142 \text{ x } (7.5 \text{ cm})^2 = 176.7 \text{ cm}^2$

Volume change = A x h = 176.7 cm^2 x 5.25 cm = 927.8 cm^3

$927.8 \text{ cm}^3 \text{ x } \frac{\text{mL}}{\text{cm}^3} \text{ x } \frac{10^{-3} \text{ L}}{\text{mL}} = 0.9278 \text{ L};$ $\quad V_2 = 1.20 \text{ L} - 0.93 \text{ L} = 0.27 \text{ L}$

$$P_2 = \frac{P_1 V_1}{V_2} = \frac{1.00 \text{ atm x } 1.20 \text{ L}}{0.27 \text{ L}} = 4.4 \text{ atm}$$

43. $\frac{V_1}{T_1} = \frac{V_2}{T_2}$ $\quad V_2 = \frac{V_1 T_2}{T_1} = 154 \text{ mL x } \frac{(10 + 273)\text{K}}{(100 + 273)\text{K}} = 117 \text{ mL}$

45. $T_2 = \frac{T_1 V_2}{V_1} = (305 + 273)\text{K x } \frac{425 \text{ mL}}{567 \text{ mL}} = 433 \text{ K}$ $\quad 433 \text{ K} - 273 = 160 \text{ °C}$

47. The answer can be determined by estimation.

a. 5.0 g $H_2 \approx 2.5$ mol

b. 50 L SF_6 at STP is a little more than 2.0 mol but less than 2.5 mol.

c. 1.0 x 10 ^{24} is less than 2 mol.

(a) is the largest number of molecules.

actual calculations:

a. $5.0 \text{ g } H_2 \text{ x } \frac{\text{mol } H_2}{2.016 \text{ g } H_2} \text{ x } \frac{6.022 \text{ x } 10^{23} \text{ molecules}}{\text{mol}} = 1.5 \text{ x } 10^{24} \text{ molecules}$

b. $50 \text{ L } SF_6 \text{ x } \frac{\text{mol } SF_6}{22.4 \text{ L } SP_6} \text{ x } \frac{6.022 \text{ x } 10^{23} \text{ molecules}}{\text{mol}} = 1.3 \text{ x } 10^{24} \text{ molecules}$

c. 1.0 x 10^{24} molecules

(a) is the largest number of molecules.

49. $498 \text{ L x } \frac{\text{mol}}{22.4 \text{ L}} \text{ x } \frac{20.18 \text{ g}}{\text{mol}} = 449 \text{ g Ne}$

51. $\frac{P_1}{T_1} = \frac{P_2}{T_2}$ $\quad P_2 = P_1 \text{ x } \frac{T_2}{T_1} = 721 \text{ torr x } \frac{(755 + 273)\text{K}}{(25 + 273)\text{K}} = 2.49 \text{ x } 10^3 \text{ torr}$

53. 1 mol at 273 K and 1 atm is 22.4 L.

$$\frac{P_1 V_1}{T_1} = \frac{P_2 V_2}{T_2}, \quad P_1 = P_2, V_2 = V_1 \text{ x } \frac{T_2}{T_1} = \frac{22.4 \text{ L x } (25 + 273)\text{K}}{273\text{K}} = 24.5 \text{ L}$$

55. $V_2 = V_1 \text{ x } \frac{P_1 T_2}{P_2 T_1} = 2.53 \text{ m}^3 \text{ x } \frac{191 \text{ torr}}{1142 \text{ torr}} \text{ x } \frac{(25 + 273)\text{K}}{(-15 + 273)\text{K}} = 0.489 \text{ m}^3$

57. a. $V = \frac{nRT}{P} = \frac{1.12 \text{ mol x } \frac{0.08206 \text{ L atm}}{\text{K mol}} \text{ x } (62 + 273)\text{K}}{1.38 \text{ atm}} = 22.3 \text{ L}$

b. $$V = \frac{nRT}{P} = \frac{6.00 \times 10^{-3}\ \text{mol} \times \frac{0.08206\ \text{L atm}}{\text{K mol}} \times (31 + 273)\text{K}}{661\ \text{mmHg} \times \frac{\text{atm}}{760\ \text{mmHg}}} \times \frac{\text{mL}}{10^{-3}\ \text{L}} = 172\ \text{mL}$$

59. a. $$P = \frac{nRT}{V} = \frac{4.64\ \text{mol} \times \frac{0.08206\ \text{L atm}}{\text{K mol}} \times (29 + 273)\text{K}}{3.96\ \text{L}} = 29.0\ \text{atm}$$

b. $$P = \frac{nRT}{\text{V}} = \frac{0.0108\ \text{mol} \times \frac{0.08206\ \text{L atm}}{\text{K mol}} \times (37 + 273)\text{K}}{0.265\ \text{L}} = 1.037\ \text{atm}$$

$$P = 1.037\ \text{atm} \times \frac{760\ \text{mmHg}}{\text{atm}} = 788\ \text{mmHg}$$

61. a. $$m = \frac{PV\mathcal{M}}{RT} = \frac{698\ \text{torr} \times \frac{\text{atm}}{760\ \text{torr}} \times 2.22\ \text{L} \times \frac{83.80\ \text{g}}{\text{mol}}}{\frac{0.08206\ \text{L atm}}{\text{K mol}} \times (45 + 273)\text{K}} = 6.55\ \text{g Kr}$$

b. $$m = \frac{PV\mathcal{M}}{RT} = \frac{784\ \text{torr} \times \frac{\text{atm}}{760\ \text{torr}} \times 7.45\ \text{mL} \times \frac{10^{-3}\ \text{L}}{\text{mL}} \times \frac{28.01\ \text{g}}{\text{mol}}}{\frac{0.08206\ \text{L atm}}{\text{K mol}} \times (36 + 273)\text{K}} \times \frac{\text{mg}}{10^{-3}\ \text{g}}$$

$$= 8.49\ \text{mg CO}$$

63. $$n = \frac{PV}{RT} = \frac{1.02\ \text{atm} \times 1.70 \times 10^{10}\ \text{L}}{\frac{0.08206\ \text{L atm}}{\text{K mol}} \times (18 + 273)\text{K}} = 7.26 \times 10^{8}\ \text{mol}$$

65. $$\mathcal{M} = \frac{mRT}{PV} = 0.549\ \text{g} \times \frac{\frac{0.08206\ \text{L atm}}{\text{K mol}} \times (24 + 273)\text{K}}{747\ \text{mmHg} \times \frac{\text{atm}}{760\ \text{mmHg}} \times 211\ \text{mL} \times \frac{10^{-3}\ \text{L}}{\text{mL}}} = 64.5\ \text{g/mol}$$

Molecular weight = 64.5 u

67. $$\mathcal{M} = \frac{mRT}{PV} = 0.625\ \text{g} \times \frac{\frac{0.08206\ \text{L atm}}{\text{K mol}} \times (98 + 273)\text{K}}{756\ \text{torr} \times \frac{\text{atm}}{760\ \text{torr}} \times 125\ \text{mL} \times \frac{10^{-3}\ \text{L}}{\text{mL}}} = 153\ \text{g/mol}$$

Molecular weight = 153 u

69. a. $$d = \frac{m}{V} = \frac{P\mathcal{M}}{RT} = \frac{1.00 \text{ atm} \times \frac{28.01 \text{ g}}{\text{mol}}}{\frac{0.08206 \text{ L atm}}{\text{K mol}} \times 273 \text{ K}} = 1.25 \text{ g/L}$$

b. $$d = \frac{P\mathcal{M}}{RT} = \frac{1.26 \text{ atm} \times \frac{39.95 \text{ g}}{\text{mol}}}{\frac{0.08206 \text{ L atm}}{\text{K mol}} \times (325 + 273)\text{K}} = 1.03 \text{ g/L}$$

71. $$T = \frac{P\mathcal{M}}{Rd} = \frac{0.982 \text{ atm} \times \frac{32.00 \text{ g}}{\text{mol}}}{\frac{0.08206 \text{ L atm}}{\text{K mol}} \times \frac{1.05 \text{ g}}{\text{L}}} = 365 \text{ K}$$ $365 \text{ K} - 273 = 92\ ^\circ\text{C}$

73. $$\mathcal{M} = \frac{dRT}{P} = \frac{\frac{2.57 \text{ g}}{\text{L}} \times \frac{0.08206 \text{ L atm}}{\text{K mol}} \times (25 + 273)\text{K}}{745 \text{ torr} \times \frac{\text{atm}}{760 \text{ torr}}} = 64.1 \text{ g/mol}$$

75. $$1.15 \text{ L } SO_2 \times \frac{2 \text{ mol } SO_3}{2 \text{ mol } SO_2} = 1.15 \text{ L } SO_3$$

$$0.65 \text{ L } O_2 \times \frac{2 \text{ mol } SO_3}{\text{mol } O_2} = 1.30 \text{ L } SO_3$$

SO_2 limiting 1.15 L SO_3 is produced

77. $$1.00 \times 10^3 \text{ kg} \times \frac{10^3 \text{ g}}{\text{kg}} \times \frac{\text{mol } CaCO_3}{100.09 \text{ g}} \times \frac{\text{mol } CO_2}{\text{mol } CaCO_3} \times \frac{0.08206 \text{ L atm}}{\text{K mol}} \times \frac{760 \text{ mmHg}}{\text{atm}}$$

$$\times \frac{(273.2 + 22.5)\text{K}}{743.5 \text{ mmHg}} = 2.48 \times 10^5 \text{ L } CO_2.$$

79. $$\frac{PV}{RT} = n$$

$$\frac{758 \text{ torr} \times \frac{\text{atm}}{760 \text{ torr}} \times 28.50 \text{ mL} \times \frac{10^{-3} \text{ L}}{\text{mL}}}{\frac{0.08206 \text{ L atm}}{\text{K mol}} \times (26 + 273)\text{K}} \times \frac{\text{mol mg}}{\text{mol } H_2} \times \frac{24.31 \text{ g Mg}}{\text{mol Mg}} \times \frac{\text{mg}}{10^{-3} \text{ g}}$$
$$= 28.2 \text{ mg Mg}$$

81. a. $0.354 \text{ g Ar} \times \frac{\text{mol}}{39.95 \text{ g}} = 0.008861 \text{ mol}$

$0.0521 \text{ g Ne} \times \frac{\text{mol}}{20.18 \text{ g}} = 0.002582 \text{ mol}$

$0.00419 \text{ g Kr} \times \frac{\text{mol}}{83.80} = \underline{0.0000500 \text{ mol}}$

0.011493 mol

$\chi_{Ar} = \frac{0.008861 \text{ mol Ar}}{0.011493 \text{ mol}} = 0.771$

$\chi_{Ne} = \frac{0.002582 \text{ mol Ne}}{0.011493 \text{ mol}} = 0.225$

$\chi_{Kr} = \frac{0.0000500 \text{ mol Kr}}{0.011493 \text{ mol}} = 0.00435$

b. $1.98 \text{ g } N_2 \times \frac{\text{mol}}{28.01 \text{ g}} = 0.07069 \text{ mol} \times \frac{1}{0.08338 \text{ mol}} = 0.848 = \chi_{N_2}$

$0.390 \text{ g } O_2 \times \frac{\text{mol}}{32.00 \text{ g}} = 0.01219 \text{ mol} \times \frac{1}{0.08338 \text{ mol}} = 0.146 = \chi_{O_2}$

$0.0201 \text{ g Ar} \times \frac{\text{mol}}{39.95 \text{ g}} = \underline{0.00050} \text{ mol} \times \frac{1}{0.08338 \text{ mol}} = 0.00603 = \chi_{Ar}$

0.08338 mol

83. Volume fraction equals mole fraction.

$P_{CO_2} = 0.44 \times 818 \text{ torr} = 3.6 \times 10^2 \text{ torr}$

$P_{H_2} = 0.38 \times 818 \text{ torr} = 3.1 \times 10^2 \text{ torr}$

$P_{N_2} = 0.17 \times 818 \text{ torr} = 1.4 \times 10^2 \text{ torr}$

$P_{O_2} = 0.013 \times 818 \text{ torr} = 11 \text{ torr}$

$P_{CH_4} = 0.00003 \times 818 \text{ torr} = 2 \times 10^{-2} \text{ torr}$

85. $P_{total} = P_{O_2} + P_{H_2O}$

$P_{O_2} = 742 \text{ torr} - 32 \text{ torr} = 710 \text{ torr}$

87. $P_{O_2} = 743 \text{ torr} - 19 \text{ torr} = 724 \text{ torr}$

$$n = \frac{PV}{RT} = \frac{724 \text{ torr} \times \frac{\text{atm}}{760 \text{ torr}} \times 122 \text{ mL} \times \frac{10^{-3}\text{ L}}{\text{mL}}}{\frac{0.08206 \text{ L atm}}{\text{K mol}} \times (21 + 273)\text{K}}$$

$$n = 4.82 \times 10^{-3} \text{ mol } O_2$$

$$4.82 \times 10^{-3} \text{ mol } O_2 \times \frac{\text{mol } C_6H_{12}O_6}{6 \text{ mol } O_2} \times \frac{180.16 \text{ g}}{\text{mol}} = 0.145 \text{ g } C_6H_{12}O_6$$

89. $$\frac{r_{H_2}}{r_{He}} = \sqrt{\frac{4.003}{2.016}} = 1.41$$

H_2 will diffuse somewhat faster than He, so b, with a ratio of H_2 to He molecules of 6:4 = 1.5, is the answer.

91. a. $r \; \alpha \; \frac{1}{t}$

$$\frac{r_{N_2}}{r_X} = \frac{\frac{1}{44 \text{ s}}}{\frac{1}{75 \text{ s}}} = \sqrt{\frac{\mathcal{M}_X}{\mathcal{M}_{N_2}}} = \sqrt{\frac{\mathcal{M}_X}{28.01 \text{ g/mol}}}$$

$$1.70 = \sqrt{\frac{\mathcal{M}_X}{28.01 \text{ g/mol}}}$$

$$2.91 = \frac{\mathcal{M}_X}{28.01 \text{ g/mol}}$$

$\mathcal{M}_X = 81.5$ g/mol. Molecular weight = 81 u.

b. $$\frac{r_{N_2}}{r_x} = \frac{\frac{1}{44 \text{ s}}}{\frac{1}{42 \text{ s}}} = \sqrt{\frac{\mathcal{M}_X}{28.01 \text{ g/mol}}}$$

$$0.955 = \sqrt{\frac{\mathcal{M}_X}{28.01 \text{ g/mol}}}$$

$\mathcal{M}_X = 0.911 \times 28.01$ g/mol = 26 g/mol. Molecular weight = 26 u

93. $P_{Ne} = 3.00 \text{ atm} - \left(735 \text{ mmHg x } \frac{\text{atm}}{760 \text{ mmHg}}\right) = 2.03 \text{ atm}$

$$m = \frac{PVM}{RT} = \frac{2.03 \text{ atm x } 1.83 \text{ L x } 20.18 \text{ g/mol}}{\frac{0.08206 \text{ L atm}}{\text{K mol}} \text{ x } (18 + 273)\text{K}}$$

$m = 3.14$ g Ne

95. H_2 $\quad 0.0080 \text{ g x } \frac{\text{mol}}{2.016 \text{ g}} = 0.0040 \text{ mol}$

N_2 $\quad 0.1112 \text{ g x } \frac{\text{mol}}{28.014 \text{ g}} = 0.003969 \text{ mol}$

O_2 $\quad 0.1281 \text{ g x } \frac{\text{mol}}{31.999 \text{ g}} = 0.004003 \text{ mol}$

CO_2 $\quad 0.1770 \text{ g x } \frac{\text{mol}}{44.009 \text{ g}} = 0.004022 \text{ mol}$

C_4H_{10} $\quad 0.2320 \text{ g x } \frac{\text{mol}}{58.124 \text{ g}} = 0.003991 \text{ mol}$

CCl_2F_2 $\quad 0.4824 \text{ g x } \frac{\text{mol}}{120.91 \text{ g}} = 0.003990 \text{ mol}$

These data are consistent with Avogadro's hypothesis. The amount of each gas is within 1% of 0.00400 mol, strongly suggesting that equal volumes of different gases, compared at the same temperature and pressure, have equal numbers of molecules.

97. Calculate the empirical formula.

$$85.63 \text{ g C x } \frac{\text{mol C}}{12.011 \text{ g C}} = 7.1293 \text{ mol C x } \frac{1}{7.1293 \text{ mol C}} = 1.00$$

$$14.37 \text{ g H x } \frac{\text{mol H}}{1.0079 \text{ g C}} = 14.2574 \text{ mol H x } \frac{1}{7.1293 \text{ mol C}} = 2.00$$

CH_2 $\quad 12.011 + 2(1.0079) = \frac{14.027 \text{ u}}{\text{formula unit}}$

empirical formula

$$\mathcal{M} = \frac{dRT}{P} = \frac{\frac{1.69 \text{ g}}{\text{L}} \text{ x } \frac{0.08206 \text{ L atm}}{\text{K mol}} \text{ x } (24 + 273)\text{K}}{743 \text{ mmHg x } \frac{\text{atm}}{760 \text{ mmHg}}} = 42.13 \text{ g/mol}$$

$$\frac{42.13 \text{ u}}{\text{molecule}} \text{ x } \frac{\text{formula unit}}{14.027 \text{ u}} = 3.00 \frac{\text{formula unit}}{\text{molecule}}$$

C_3H_6

99. Use the freon data to calculate the glass volume.

192.8273 g freon + glass	45.2217 g B + glass
- 45.0143 g glass	45.0143 g glass
147.8130 g freon	0.2074 g B

$$147.8130 \text{ g x } \frac{\text{mL}}{1.576 \text{ g}} = 93.790 \text{ mL volume glass}$$

$$\mathcal{M} = \frac{mRT}{PV} = \frac{0.2074 \text{ g x } \frac{0.082057 \text{ L atm}}{\text{K mol}} \text{ x } (21.48 + 273.15)\text{K}}{751.2 \text{ mmHg x } \frac{\text{atm}}{760 \text{ mmHg}} \text{ x } 93.790 \text{ mL x } \frac{10^{-3} \text{ L}}{\text{mL}}}$$

$\mathcal{M} = 54.09$ g/mol. Molecular weight = 54.09 u

101. $2\ C_8H_{18} + 25\ O_2 \longrightarrow 16\ CO_2 + 18\ H_2O$

$$265 \text{ mi x } \frac{\text{gal}}{31.2 \text{ mi}} \text{ x } \frac{3.785 \text{ L}}{\text{gal}} \text{ x } \frac{\text{mL}}{10^{-3} \text{ L}} \text{ x } \frac{0.71 \text{ g}}{\text{mL}} \text{ x } \frac{\text{mol}}{114 \text{ g}} \text{ x } \frac{16 \text{ mol } CO_2}{2 \text{ mol } C_8H_{18}} \text{ x}$$

$$\frac{0.0821 \text{ L atm}}{\text{K mol}} \text{ x } \frac{(28 + 273)\text{K}}{732 \text{ mmHg x } \frac{\text{atm}}{760 \text{ mmHg}}} = 4.1 \text{ x } 10^4 \text{ L } CO_2$$

103. $$U_{rms} = \sqrt{\frac{3\,RT}{\text{M}}}$$

$$U_{rms} = \sqrt{\frac{3 \text{ x } \frac{8.3145 \text{ kg}\cdot\text{m}^2\cdot\text{s}^{-2}}{\text{mol K}} \text{ x } (27 + 273)\text{K}}{0.06407 \text{ kg/mol}}} = 342 \text{ m/s}$$

105. $$V = \frac{nRT}{P} = \frac{1.00 \text{ mole } \frac{0.08206 \text{ L atm}}{\text{K mol}} (25 + 273)\text{K}}{755 \text{ mmHg x } \frac{\text{atm}}{760 \text{ mmHg}}} = 24.6 \text{ L}$$

$$24.6 \text{ L x } \frac{\text{mL}}{10^{-3} \text{ L}} \text{ x } \frac{\text{s}}{80 \text{ mL}} = 3.1 \text{ x } 10^2 \text{ s}$$

Chapter 5

Thermochemistry

Exercises

5.1 $\Delta E = q + w = (-567\ J) + (+89\ J) = -478\ J$

5.2 $\Delta H_{rxn} = \frac{1}{2}(285.4\ kJ) = 142.7\ kJ$

5.3 $\Delta H_{rxn} = -\frac{1}{4}(595.5\ kJ) = -148.9\ kJ$

5.4 $\Delta H_{rxn} = 12.8\ g\ H_2 \times \frac{mol\ H_2}{2.016\ g\ H_2} \times \frac{-184.6\ kJ}{mol} = -1.17 \times 10^3\ kJ$

5.5 $-1.00 \times 10^6\ kJ \times \frac{mol}{-890.3\ kJ} \times \frac{0.08206\ L\ atm}{K\ mol} \times \frac{(273 + 25)K}{745\ torr \times \frac{atm}{760\ torr}} =$

$2.80 \times 10^4\ L\ CH_4$

5.6 $C = \frac{q}{\Delta T} = \frac{911\ J}{100\ °C - 15\ °C} = 11\ J/°C$

5.7 $q = m$ times specific heat times ΔT

$q = 814\ g \times \frac{4.182\ J}{g\ °C} \times (100.0\ °C - 18.0\ °C) \times \frac{1.00\ cal}{4.184\ J} = 6.67 \times 10^4\ cal$

$q = 6.67 \times 10^4\ cal \times \frac{kcal}{10^3\ cal} = 66.7\ kcal$

5.8 $T_f - T_i = \frac{q}{m \times \text{specific heat}}$

$T_f = 22.5\ °C + \frac{4.22\ kJ \times \frac{10^3\ J}{kJ}}{454\ g \times 0.13\ J/g\ °C}$

$T_f = 94.0\ °C$

5.9 $q_{Cu} = q_{H_2O}$

$m_{Cu} \times \text{specific heat}_{Cu} \times \Delta T = m_{H_2O} \times \text{specific heat} \times \Delta T$

$m_{Cu} \times \frac{0.385\ J}{g\ °C} \times \Delta T = 145\ g\ H_2O \times \frac{4.182\ J}{g\ °C} \times \Delta T$ (ΔT cancels)

$m_{Cu} = 1.57 \times 10^3\ g\ Cu$

5.10 $q_{Ir} = -q_{H_2O}$

$m_{Ir} \text{ x specific heat x } \Delta T = -m_{H_2O} \text{ x specific heat}_{H_2O} \text{ x } \Delta T$

$23.9 \text{ g Ir x specific heat}_{Ir} \text{ x } (22.6\ ^\circ\text{C} - 89.7\ ^\circ\text{C}) = -20.0 \text{ g } H_2O \text{ x } \frac{4.182 \text{ J}}{\text{g }^\circ\text{C}} \text{ x}$

$(22.6\ ^\circ\text{C} - 20.1\ ^\circ\text{C})$
specific heat$_{Ir}$ = 0.13 J /g °C

5.11 $100.0 \text{ mL x } 0.500 \text{ M x } \frac{10^{-3} \text{ L}}{\text{mL}} = 0.0500 \text{ mol product}$

$\Delta H = -q_{\text{calorim}} = m \text{ x specific heat x } \Delta T$

$\Delta H = -200.0 \text{ mL x } \frac{1.00 \text{ g}}{\text{mL}} \text{ x } \frac{4.182 \text{ J}}{\text{g }^\circ\text{C}} \text{ x } (23.65\ ^\circ\text{C} - 20.29\ ^\circ\text{C})$

$\Delta H = \frac{-2.81 \text{ x } 10^3 \text{ J}}{0.0500 \text{ mol}} = -5.62 \text{ x } 10^4 \text{ J/mol} = -56.2 \text{ kJ/mol}$

5.12 $\Delta H = -q_{\text{calorim}} = \text{heat capacity x } \Delta T$

$\Delta H = \frac{-5.15 \text{ kJ}}{^\circ\text{C}} \text{ x } (28.05\ ^\circ\text{C} - 25.00\ ^\circ\text{C}) \text{ x } \frac{10^3 \text{ J}}{\text{kJ}} = -1.57 \text{ x } 10^4 \text{ J}$

$\Delta\text{H} = \frac{-1.57 \text{ x } 10^4 \text{ J}}{0.480 \text{ g C}} \text{ x } \frac{12.01 \text{ g C}}{\text{mol C}} = 3.93 \text{ x } 10^5 \text{ J/mol C} = 393 \text{ kJ/mol C}$

5.13 $0.8082 \text{ g x } \frac{\text{mol}}{180.16 \text{ g}} \text{ x } \frac{-2803 \text{ kJ}}{\text{mol}} = -12.57 \text{ kJ}$

$\text{heat capacity} = \frac{-q_{\text{calorim}}}{\Delta T} = \frac{12.57 \text{ kJ}}{(27.21 - 25.11)^\circ\text{C}} = \frac{5.99 \text{ kJ}}{^\circ\text{C}}$

5.14

$2 \text{ C (graphite)} + 2\ O_2 \text{ (g)} \to 2\ CO_2 \text{ (g)}$	2 x - 393.5 kJ
$2\ CO_2 \text{ (g)} + 2\ H_2O \text{ (l)} \to 3\ O_2 \text{ (g)} + C_2H_4 \text{ (g)}$	- (-1410.9 kJ)
$2\ H_2 \text{ (g)} + O_2 \text{ (g)} \to 2\ H_2O \text{ (l)}$	- 571.6 kJ
$2 \text{ C (graphite)} + 2\ H_2 \text{ (g)} \to C_2H_4 \text{ (g)}$ $\quad \Delta H =$	52.3 kJ

5.15

$C_2H_4 \text{ (g)} + 3\ O_2 \text{ (g)} \to 2\ CO_2 \text{ (g)} + 2\ H_2O \text{ (l)}$	- 1410.9 kJ
$3\ H_2O \text{ (l)} + 2\ CO_2 \text{ (g)} \to \frac{7}{2} O_2 \text{ (g)} + C_2H_6 \text{ (g)}$	- (-1559.7 kJ)
$H_2 \text{ (g)} + \frac{1}{2} O_2 \text{ (g)} \to H_2O \text{ (l)}$	$\frac{1}{2}$(-571.6 kJ)
$C_2H_4 \text{ (g)} + H_2 \text{ (g)} \to C_2H_6 \text{ (g)}$	ΔH = - 137 kJ

5.16 $\Delta H^\circ = \Sigma\nu_p\Delta H^\circ_f \text{ (products)} - \Sigma\nu_\rho\Delta H^\circ_f \text{ (reactants)}$

$\Delta H^\circ = (1 \text{ mol } CH_3CH_2OH \text{ x } -277.7 \text{ kJ/mol}) - (1 \text{ mol } H_2O \text{ x } -285.8 \text{ kJ/mol} + 1 \text{ mol } C_2H_4 \text{ x } 52.26 \text{ kJ/mol}) = -44.2 \text{ kJ}$

5.17 $\Delta H^\circ = \Sigma\nu_p\Delta H^\circ_f$ (products) - $\Sigma\nu_r\Delta H^\circ_f$ (reactants)
-2523 kJ = (4 mol CO_2 x -393.5 kJ/mol + 2 mol H_2O x -285.8 kJ /mol
+ 1 mol SO_2 x -296.8 kJ/mol) - 1(mol C_4H_4S x $\Delta \overset{\circ}{H_f}$ - 6 mol O_2 x 0)
ΔH°_f = 80.6 kJ /mol

Estimation Exercises

5.1 $\Delta T = \dfrac{q}{m \text{ x specific heat}}$
The smallest specific heat will produce the largest ΔT. Silver will be the hottest; iron second hottest.

5.2 CH_4 - 890 kJ

C_2H_6 $\dfrac{-1560 \text{ kJ}}{2}$ = -780 kJ C_6H_{14} $\dfrac{-4163 \text{ kJ}}{6}$ = -693 kJ

C_3H_8 $\dfrac{-2220 \text{ kJ}}{3}$ = -740 kJ C_7H_{16} $\dfrac{-4811 \text{ kJ}}{7}$ = -687 kJ

C_4H_{10} $\dfrac{-2879 \text{ kJ}}{4}$ = -720 kJ C_8H_{18} $\dfrac{-5450 \text{ kJ}}{8}$ = -681 kJ

C_5H_{12} $\dfrac{-3536 \text{ kJ}}{5}$ = -707 kJ

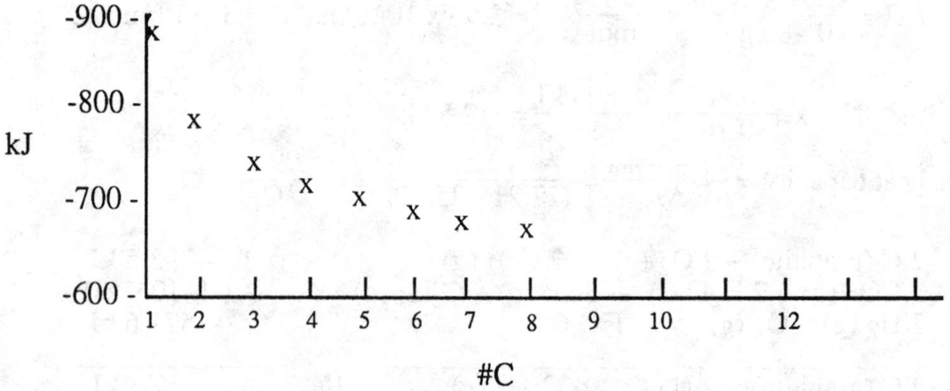

The curve is almost linear for 6 through 8 C. Extending that line produces
-687 - (-693) = 6; -681 - (-687) = 6; -681 +12 = -669; -669 + 12 = -657

$\Delta H\ (C_{10}H_{22}) = \dfrac{-669 \text{ kJ}}{C}$ x 10C ≈ -6700 kJ

$\Delta H\ (C_{12}H_{26}) = \dfrac{-657 \text{ kJ}}{C}$ x 12C ≈ -7900 kJ

Conceptual Exercises

5.1 $q_{100} = -q_{200}$

m x specific heat x ΔT = $-m$ x specific heat x ΔT

100 mL x (T_f - 20 °C) = -200 mL x (T_f - 80 °C)

100 T_f - 2000 °C = -200 T_f + 16000 °C

300 T_f = 18000 °C

T_f = 60 °C

OR

The heat exchange produces twice the temperature increase in the 100-mL sample, 2x, as tthe temperature decrease, x, in the 200-mL sample.

x + 2x = 60 °   x = 20 °

The final temperature is 20 °C + (2 x 20 °) = 80 °C - 20 °C = 60 °C

5.2 The enthalpy of formation of CH_3OH is 39 kJ/mol greater than that of CH_3CH_2OH, but the enthalpies of the products of the combustion of CH_3CH_2OH are much lower than of the combustion of CH_3OH [by one mole each of H_2O(l) andCO_2(g)]. CH_3CH_2OH has the greater negative heat of combustion (recall Figure 5.14).

Review Questions

13. $CaCO_3$ (s) —> CaO (s) + CO_2 (g)

endothermic reaction

$$q = \Delta H = \frac{17.8 \text{ kJ}}{0.100 \text{ mol}} = \frac{178 \text{ kJ}}{\text{mol } CaCO_3}$$

15. $\Delta H_{rxn} = -\frac{1}{2}\Delta H = -\frac{1}{2}(-92.22 \text{ kJ}) = +46.11 \text{ kJ}$

17. The heat capacity is the amount of heat required to change the temperature of the system by one degree C. It is often expressed in joules per °C or in joules per kelvin.

19. A measured amount of the substance absorbing a measured amount of heat will produce a temperature change. That change will allow the specific heat to be calculated with the equation q = mass x specific heat x ΔT.

21. Reactants are placed in a calorimeter that can be sealed to maintain a constant volume. The calorimeter is immersed in water. The reaction occurs, often initiated by a spark. The change of temperature of the water allows calculation of the heat of reaction. A reaction of known heat of reaction is used to determine the heat capacity of the water and calorimeter. Combustion reactions are commonly studied in bomb calorimeters.

23. A substance is in its standard state when it is a pure material at 1 atm pressure. For a gas, it is the gas acting as a hypothetical ideal gas at 1 atm.

25. 2 Fe (s) + $\frac{3}{2}$ O_2 (g) —> Fe_2O_3 (s)

27. The relevant equation is $\Delta H^\circ = \Sigma \nu_p \Delta H_f^\circ$(products) - $\Sigma \nu_r \Delta H_f^\circ$(reactants). Standard heats of reaction, ΔH°, are calculated by substituting ΔH_f° values on the right side of the equation. To determine an unknown ΔH_f°, a known ΔH° is substituted on the left side of the equation and known ΔH_f° values are substituted for all the terms on the right side, except for the unknown.

29. The food Calorie is 1000 times larger than the ordinary calorie used in scientific work: 1 Cal = 1000 cal.

Problems

31. $\Delta E = 455 \text{ J} - 325 \text{ J} = 130 \text{ J}$

33. 58 cal of work must be done on the system. $58 \text{ cal} \times \frac{4.184 \text{ J}}{\text{cal}} = 2.4 \times 10^2 \text{ J}$

35. Endothermic $q = \frac{134 \text{ kJ}}{10.0 \text{ g}} \times \frac{18.02 \text{ g}}{\text{mol}} = 241 \text{ kJ/mol}$

$q_p = \Delta H$ At constant pressure; there is pressure-volume work that is included in q_p.

37. $CaCO_3 \text{ (s)} \longrightarrow CaO \text{ (s)} + CO_2 \text{ (g)}$ $\Delta H = \frac{8.90 \text{ kJ}}{0.0500 \text{ mol}} = \frac{178 \text{ kJ}}{\text{mol}}$

39. a. $\Delta H^\circ_{rxn} = \frac{1}{3}(-46 \text{ kJ}) = -15 \text{ kJ}$

b. $\Delta H^\circ_{rxn} = -\frac{1}{2}(-46 \text{ kJ}) = 23 \text{ kJ}$

41. $0.333 \text{ mol} \times \frac{-65.2 \text{ kJ}}{\text{mol}} = -21.7 \text{ kJ}$

43. $-429 \text{ kJ} \times \frac{3 \text{ mol CO}}{-24.8 \text{ kJ}} \times \frac{28.01 \text{ g}}{\text{mol}} \times \frac{\text{kg}}{10^3 \text{ g}} = 1.45 \text{ kg CO}$

45. $q = C\Delta T$

$C = \frac{112 \text{ J}}{45\,^\circ\text{C} - 18\,^\circ\text{C}} = 4.1 \text{ J/}^\circ\text{C}$

47. a. $q = 20.0 \text{ g} \times \frac{4.182 \text{ J}}{\text{g}\,^\circ\text{C}} \times (96.0\,^\circ\text{C} - 20.0\,^\circ\text{C}) \times \frac{\text{kJ}}{10^3 \text{ J}} = 6.36 \text{ kJ}$

b. $q = 120.0 \text{ g} \times \frac{2.46 \text{ J}}{\text{g}\,^\circ\text{C}} \times (44.5\,^\circ\text{C} - (-10.5\,^\circ\text{C})) \times \frac{\text{kJ}}{10^3 \text{ J}} = 16.2 \text{ kJ}$

49. $\Delta T = \dfrac{q}{m \text{ x specific heat}} = \dfrac{93.5\text{ J}}{48.7\text{ g x }\dfrac{0.128\text{ J}}{\text{g °C}}} = 15.0\text{ °C}$

$t_f = \Delta T + t_i = 15.0\text{ °C} + 27.0\text{ °C} = 42.0\text{ °C}$

51. $-q_{\text{metal}} = q_{\text{water}}$

$10.25\text{ g x specific heat x }(22.03\text{ °C} - 99.10\text{ °C}) = q_{\text{metal}}$

$20.0\text{ g x }\dfrac{4.182\text{ J}}{\text{g°C}}\text{ x }(22.03\text{ °C} - 18.51\text{ °C}) = q_{\text{water}}$

$q_{\text{water}} = 294.4\text{ J} = -q_{\text{metal}} = -(-790.0\text{ g°C x specific heat})$

specific heat = 0.373 J/g°C

53. $q_{Fe} = -q_{H_2O}$

$m_{Fe}\text{ sp. ht.}_{(Fe)}\Delta T = -m_{H_2O}\text{ sp. ht.}_{(H_2O)}\Delta T$

$1.35\text{ kg x }\dfrac{10^3\text{ g}}{\text{kg}}\text{ x }\dfrac{0.449\text{ J}}{\text{g °C}}\text{ x }(39.6 - t_i)\text{°C} = -0.817\text{ kg x }\dfrac{10^3\text{ g}}{\text{kg}}\text{ x }\dfrac{4.182\text{ J}}{\text{g °C}}\text{ x }(39.6 - 23.3)\text{°C}$

$2.400 \text{ x } 10^4\text{ J} - \dfrac{606.2\text{ J}}{\text{°C}}t_i = -5.569 \text{ x } 10^4\text{ J}$

$\dfrac{606.2\text{ J }t_i}{\text{°C}} = 5.569 \text{ x } 10^4\text{ J} + 2.400 \text{ x } 10^4\text{ J}$

$t_i = \dfrac{7.969 \text{ x } 10^4\text{ J}}{606.2\text{ J/°C}} = 131\text{ °C}$

55. $q_{\text{calorim}} = \text{mass x specific heat x }\Delta T$

$q_{\text{calorim}} = 1000\text{ mL x }\dfrac{1.00\text{ g}}{\text{mL}}\text{ x }\dfrac{4.182\text{ J}}{\text{g°C}}\text{ x }(23.21 - 20.00)\text{°C}$

$q_{\text{calorim}} = 1.342 \text{ x } 10^4\text{ J x }\dfrac{\text{kJ}}{10^3\text{ J}} = 13.42\text{ kJ}$

$500.0\text{ mL soln x }0.500\text{ M NaOH x }\dfrac{\text{mol }H_2O}{1\text{ mol NaOH}}\text{ x }\dfrac{10^{-3}\text{ L}}{\text{mL}} = 0.2500\text{ mol}$

$\Delta H = q_{rxn} = -q_{\text{calorim}} = \dfrac{-13.42\text{ kJ}}{0.2500\text{ mol}} = -\dfrac{53.7\text{ kJ}}{\text{mol}}$

57. $q_{calorim}$ = heat capacity x ΔT

$$q_{calorim} = \frac{4.62\text{ kJ}}{°\text{C}} \times (22.28 - 20.45)°\text{C} = 8.455\text{ kJ}$$

$$q_{rxn} = -\,q_{calorim} = -\frac{8.455\text{ kJ}}{0.309\text{ g}} = -\,27.4\text{ kJ/g coal}$$

q_{rxn} is equal to the change in internal energy of the system, ΔE, which is not quite equal to the change in enthalpy, ΔH. An estimate of the difference between the two is given by $\Delta H = \Delta E + \Delta n_g RT$, where Δn_g is the difference in number of moles of gaseous reactants and products. In the combustion reaction, one mole of CO_2 (g) is produced for every mole of O_2(g) consumed: $\Delta n_g = 0$, and the difference between ΔH and ΔE is negligible.

59. q_{calorim} = heat capacity x $\Delta T = -\,q_{rxn}$

$$\text{heat capacity} = \frac{-\frac{(-\,16.5\text{ kJ})}{\text{g}} \times 2.00\text{ g}}{(25.67 - 22.83)°\text{C}} = \frac{11.6\text{ kJ}}{°\text{C}}$$

61. $2\,NO_2 \longrightarrow 2\,O_2 + N_2$ $-\,(33.2\text{ kJ})$

$N_2 + 2\,O_2 \longrightarrow N_2O_4$ 9.2 kJ

$2\,NO_2 \longrightarrow N_2O_4$ $\Delta H° = -24.0\text{ kJ}$

63. $C_3H_8 + 5\,O_2 \longrightarrow 3\,CO_2 + 4\,H_2O$ $-\,2219.9\text{ kJ}$

$3\,CO_2 \longrightarrow \frac{3}{2}\,O_2 + 3\,CO$ $-\,3 \times (-\,283.0\text{ kJ})$

$C_3H_8 + \frac{7}{2}\,O_2 \longrightarrow 3\,CO + 4\,H_2O$ $\Delta H° = -\,1370.9\text{ kJ}$

65. $\Delta H°_{rxn} = \Sigma\nu_p\Delta H°_f\,(\text{products}) - \Sigma\nu_r\Delta H°_f\,(\text{reactants})$

a. $\Delta H°_{rxn} = \Delta H°_f\,[NH_4Cl\,(s)] - \Delta H°_f\,[NH_3\,(g)] - \Delta H°_f\,[HCl\,(g)]$

$\Delta H°_{rxn} = -\,314.4\text{ kJ} - (-\,46.11\text{ kJ} - 92.31\text{ kJ}) = -\,176.0\text{ kJ}$

b. $\Delta H°_{rxn} = \Delta H°_f\,[NH_4NO_3\,(s)] - \Delta H°_f\,[NH_3\,(g)] - \Delta H°_f\,[HNO_3\,(l)]$

$\Delta H°_{rxn} = -\,365.6\text{ kJ} - (-\,46.11\text{ kJ} - 173.2\text{ kJ}) = -\,146.3\text{ kJ}$

c. $\Delta H°_{rxn} = \Delta H°_f\,[CaCl_2\,(s)] + \Delta H°_f\,[Mg\,(s)] - \Delta H°_f\,[MgCl_2\,(s)] - \Delta H°_f\,[Ca\,(s)]$

$\Delta H°_{rxn} = -\,795.8\text{ kJ} + 0 - (-\,641.3\text{ kJ} - 0) = -\,154.5\text{ kJ}$

d. $\Delta H°_{rxn} = \Delta H°_f\,[CO_2\,(g)] + \Delta H°_f\,[Fe\,(s)] - \Delta H°_f[FeO\,(s)] - \Delta H°_f\,[CO\,(g)]$

$\Delta H°_{rxn} = -\,393.5\text{ kJ} + 0\text{ kJ} - (-\,272\text{ kJ} - 110.5\text{ kJ}) = -\,11\text{ kJ}$

67. $\Delta H^\circ_{rxn} = \Sigma\nu_p \Delta H^\circ_f$ (products) - $\Sigma\nu_r \Delta H^\circ_f$ (reactants)
$\Delta H^\circ_{rxn} = 2\Delta H^\circ_f [SO_2 (g)] + 2\Delta H^\circ_f [ZnO (s)] - 2\Delta H^\circ_f [ZnS (s)] - 3\Delta H^\circ_f [O_2 (g)]$

$- 2\Delta H^\circ_f [ZnS (s)] = - 878.2$ kJ - 2 mol x - 296.8 kJ /mol - 2 mol x - 348.3 kJ/mol +
3 mol x 0 kJ /mol = 412.0 kJ

$$\Delta H^\circ_f [ZnS (s)] = \frac{412.0 \text{ kJ}}{- 2 \text{ mol}} = - 206.0 \text{ kJ /mol}$$

69. $$\frac{-23.50 \text{ kJ}}{1.050 \text{ g}} \times \frac{106.12 \text{ g}}{\text{mol}} = \frac{-2.3751 \times 10^3 \text{ kJ}}{\text{mol}}$$

$C_4H_{10}O_3 + 5\ O_2\ (g) \rightarrow 4\ CO_2\ (g) + 5\ H_2O\ (l)$
$\Delta H^\circ_{rxn} = \Sigma\nu_p \Delta H^\circ_f$ (products) - $\Sigma\nu_r \Delta H^\circ_f$ (reactants)

-2375.1 kJ /mol = 4 mol CO_2 x -393.5 kJ /mol + 5 mol H_2O x -285.8 kJ /mol

-5 mol O_2 x O - 1 mol $C_4H_{10}O_6$ ΔH°_f
ΔH°_f = -628 kJ /mol

71. a. C_8H_{18}

b. $$\frac{- 890 \text{ kJ}}{\text{mol}} \times \frac{\text{mol}}{16.04 \text{ g}} = \frac{- 55.5 \text{ kJ}}{\text{g } CH_4}$$

$$\frac{- 1560 \text{ kJ}}{\text{mol}} \times \frac{\text{mol}}{30.07 \text{ g}} = \frac{- 51.9 \text{ kJ}}{\text{g } C_2H_6}$$

$$\frac{- 2220 \text{ kJ}}{\text{mol}} \times \frac{\text{mol}}{44.09 \text{ g}} = \frac{- 50.4 \text{ kJ}}{\text{g } C_3H_8}$$

$$\frac{- 2879 \text{ kJ}}{\text{mol}} \times \frac{\text{mol}}{58.12 \text{ g}} = \frac{- 49.5 \text{ kJ}}{\text{g } C_4H_{10}}$$

$$\frac{- 3536 \text{ kJ}}{\text{mol}} \times \frac{\text{mol}}{72.15 \text{ g}} = \frac{- 49.0 \text{ kJ}}{\text{g } C_5H_{12}}$$

$$\frac{- 4163 \text{ kJ}}{\text{mol}} \times \frac{\text{mol}}{86.17 \text{ g}} = \frac{- 48.3 \text{ kJ}}{\text{g } C_6H_{14}}$$

$$\frac{- 4811 \text{ kJ}}{\text{mol}} \times \frac{\text{mol}}{100.2 \text{ g}} = \frac{- 48.0 \text{ kJ}}{\text{g } C_7H_{16}}$$

$$\frac{- 5450 \text{ kJ}}{\text{mol}} \times \frac{\text{mol}}{114.2 \text{ g}} = \frac{- 47.7 \text{ kJ}}{\text{g } C_8H_{18}}$$

CH_4

c. $$\frac{-890\text{ kJ}}{\text{mol}} \times \frac{\text{mol } CH_4}{12.01\text{ g C}} = \frac{-74.1\text{ kJ}}{\text{g C}}$$

$$\frac{-1560\text{ kJ}}{\text{mol}} \times \frac{\text{mol } C_2H_6}{24.02\text{ g C}} = \frac{-64.9\text{ kJ}}{\text{g C}}$$

$$\frac{-2220\text{ kJ}}{\text{mol}} \times \frac{\text{mol } C_3H_8}{36.03\text{ g C}} = \frac{-61.6\text{ kJ}}{\text{g C}}$$

$$\frac{-2879\text{ kJ}}{\text{mol}} \times \frac{\text{mol } C_4H_{10}}{48.04\text{ g C}} = \frac{-59.9\text{ kJ}}{\text{g C}}$$

$$\frac{-3536\text{ kJ}}{\text{mol}} \times \frac{\text{mol } C_5H_{12}}{60.05\text{ g C}} = \frac{-58.9\text{ kJ}}{\text{g C}}$$

$$\frac{-4163\text{ kJ}}{\text{mol}} \times \frac{\text{mol } C_6H_{14}}{72.06\text{ g C}} = \frac{-57.8\text{ kJ}}{\text{g C}}$$

$$\frac{-4811\text{ kJ}}{\text{mol}} \times \frac{\text{mol } C_7H_{16}}{84.07\text{ g C}} = \frac{-57.2\text{ kJ}}{\text{g C}}$$

$$\frac{-5450\text{ kJ}}{\text{mol}} \times \frac{\text{mol } C_8H_{18}}{96.08\text{ g C}} = \frac{-56.7\text{ kJ}}{\text{g C}}$$

CH_4

d. $$\frac{-890\text{ kJ}}{\text{mol}} \times \frac{\text{mol } CH_4}{\text{mol } CO_2} \times \frac{\text{mol } CO_2}{44.01\text{ g } CO_2} = \frac{-20.2\text{ kJ}}{\text{g } CO_2}$$

$$\frac{-1560\text{ kJ}}{\text{mol}} \times \frac{\text{mol } C_2H_6}{2\text{ mol } CO_2} \times \frac{\text{mol } CO_2}{44.01\text{ g } CO_2} = \frac{-17.7\text{ kJ}}{\text{g } CO_2}$$

$$\frac{-2220\text{ kJ}}{\text{mol}} \times \frac{\text{mol } C_3H_8}{3\text{ mol } CO_2} \times \frac{\text{mol } CO_2}{44.01\text{ g } CO_2} = \frac{-16.8\text{ kJ}}{\text{g } CO_2}$$

$$\frac{-2879\text{ kJ}}{\text{mol}} \times \frac{\text{mol } C_4H_{10}}{4\text{ mol } CO_2} \times \frac{\text{mol } CO_2}{44.01\text{ g } CO_2} = \frac{-16.4\text{ kJ}}{\text{g } CO_2}$$

$$\frac{-3536\text{ kJ}}{\text{mol}} \times \frac{\text{mol } C_5H_{12}}{5\text{ mol } CO_2} \times \frac{\text{mol } CO_2}{44.01\text{ g } CO_2} = \frac{-16.1\text{ kJ}}{\text{g } CO_2}$$

$$\frac{-4163\text{ kJ}}{\text{mol}} \times \frac{\text{mol } C_6H_{14}}{6\text{ mol } CO_2} \times \frac{\text{mol } CO_2}{44.01\text{ g } CO_2} = \frac{-15.8\text{ kJ}}{\text{g } CO_2}$$

$$\frac{-4811\text{ kJ}}{\text{mol}} \times \frac{\text{mol } C_7H_{16}}{7\text{ mol } CO_2} \times \frac{\text{mol } CO_2}{44.01\text{ g } CO_2} = \frac{-15.6\text{ kJ}}{\text{g } CO_2}$$

$$\frac{-5450\text{ kJ}}{\text{mol}} \times \frac{\text{mol } C_8H_{18}}{8\text{ mol } CO_2} \times \frac{\text{mol } CO_2}{44.01\text{ g } CO_2} = \frac{-15.5\text{ kJ}}{\text{g } CO_2}$$

CH_4

73. $4.8 \text{ g} \times \frac{\text{mol}}{342.3 \text{ g}} \times \frac{-5.65 \times 10^3 \text{ kJ}}{\text{mol}} \times \frac{1 \text{ kcal}}{4.184 \text{ kJ}} \times \frac{\text{Cal}}{\text{kcal}} = 18.9 \text{ Cal}$

Claim is verified.

75. a. Yes, ΔPV is changed so work is done.
b. Yes, a quantity of heat is absorbed equal to the quantity of work done.
c. $\Delta E = 0$

77. $q_{Fe} = -q_{H_2O}$

$m_{Fe}\,\text{sp.ht.(Fe)}\,\Delta T = -m_{H_2O}\,\text{sp.ht.}(H_2O)\Delta T$

$100 \text{ g} \frac{0.449 \text{ J}}{\text{g °C}} (t_f - 100)\text{°C} = -100 \text{ g} \frac{4.182 \text{ J}}{\text{g °C}} (t_f - 20)\text{°C}$

$-4.49 \times 10^3 \text{ J} + \frac{44.9 \text{ J}}{\text{°C}} t_f = 8.364 \times 10^3 \text{ J} - \frac{418.2 \text{ J}}{\text{°C}} t_f$

$\frac{463.1\text{J}}{\text{°C}} t_f = 12.85 \times 10^3 \text{ J}$

$t_f = 27.8 \text{ °C}$

79. H_2 produces no CO_2.

CH_4 $\frac{\text{mol}}{890 \text{ kJ}} \times \frac{\text{mol } CO_2}{\text{mol } CH_4} \times \frac{44.01 \text{ g } CO_2}{\text{mol } CO_2} = \frac{4.94 \times 10^{-2} \text{ g } CO_2}{\text{kJ}}$

CH_3OH $\frac{\text{mol}}{726 \text{ kJ}} \times \frac{\text{mol } CO_2}{\text{mol } CH_3OH} \times \frac{44.01 \text{ g } CO_2}{\text{mol } CO_2} = \frac{6.06 \times 10^{-2} \text{ g } CO_2}{\text{kJ}}$

C_8H_{16} $\frac{\text{mol}}{5450 \text{ kJ}} \times \frac{8 \text{ mol } CO_2}{\text{mol } C_8H_{18}} \times \frac{44.01 \text{ g } CO_2}{\text{mol } CO_2} = \frac{6.46 \times 10^{-2} \text{ g } CO_2}{\text{kJ}}$

81. $1 \text{ gal} \times \frac{3.785 \text{ L}}{\text{gal}} \times \frac{\text{mL}}{10^{-3} \text{ L}} \times \frac{0.703 \text{ g}}{\text{mL}} \times \frac{\text{mol}}{114.2 \text{ g}} \times \frac{5.45 \times 10^3 \text{ kJ}}{\text{mol}} = 1.27 \times 10^5 \text{ kJ}$

83. $\Delta T = \frac{q}{m \times \text{specific heat}}$

$852 \text{ kJ} \times \frac{10^3 \text{ J}}{\text{kJ}} \times \frac{\text{g °C}}{0.8 \text{ J}} \times \frac{1}{111.7 \text{ g Fe} + 102.0 \text{ g } Al_2O_3} = 4984 \text{ °C}$

$\Delta T = t_f - t_i = 4984 \text{ °C} = t_f - 25 \text{ °C}$

$t_f \approx 5000 \text{ °C}$ This is far above the melting point of iron (1530 °C).

85. $q_{naphthalene} = -q_{calorimeter}$ = -heat capacity x ΔT

$$\text{heat capacity} = \frac{\frac{5153.5\text{ kJ}}{\text{mol}} \times \frac{\text{mol}}{128.16\text{ g}} \times 1.108\text{ g}}{5.92\text{ °C}}$$

heat capacity = 7.526 kJ /°C

$q_{thymol} = -q_{calorimeter}$ = -heat capacity x ΔT = -7.526 kJ x 6.74 °C

q thymol = -50.73 kJ

$$\Delta H_{comb}(\text{thymol}) = \frac{q}{\text{mol}} = \frac{-50.73\text{ kJ}}{1.351\text{ g} \times \frac{\text{mol}}{150.2\text{ g}}} = -5.64 \times \frac{10^3\text{ kJ}}{\text{mol}}$$

87. $$\text{heat generated} = 500\text{ mL} \times \frac{10^{-3}\text{ L}}{\text{mL}} \times 6.0\text{ M} \times \frac{-42\text{ kJ}}{\text{mol}} \times \frac{10^3\text{ J}}{\text{kJ}} = -1.26 \times 10^5\text{ J}$$

$q = -m\text{sp.ht.}\Delta T$

$$-1.26 \times 10^5\text{ J} = -500\text{ g} \times \frac{4.18\text{ J}}{\text{g °C}} \times (t_f - 25\text{ °C})$$

t_f = 85°C

Assuming specific heat of solution is the same as water and that 500 mL equals 500 g and the initial temperature is room temperature of 25°C.

89.

$BrCl(g) \longrightarrow \frac{1}{2}Br_2(l) + \frac{1}{2}Cl_2(g)$	$-\frac{1}{2} \times 29.2$ kJ
$\frac{1}{2}Br_2(l) \longrightarrow \frac{1}{2}Br_2(g)$	$\frac{1}{2} \times 30.91$ kJ
$\frac{1}{2}Br_2(g) \longrightarrow Br(g)$	$\frac{1}{2} \times 192.9$ kJ
$\frac{1}{2}Cl_2(g) \longrightarrow Cl(g)$	$\frac{1}{2} \times 243.4$ kJ
$BrCl(g) \longrightarrow Br(g) + Cl(g)$	$\Delta H°$ = 219.0 kJ

91. Because ΔH_f° [$H_2O(l)$] is more negative than ΔH_f° [$H_2O(g)$], reaction (a) should liberate the greater quantity of heat, by 572 kJ/mol $C_{12}H_{26}$.

$$\left(\frac{285.8\text{ kJ}}{\text{mol}} - \frac{241.8\text{ kJ}}{\text{mol}}\right) \times \frac{26\text{ mol } H_2O}{2\text{ mol } C_{12}H_{26}} = \frac{572\text{ kJ}}{\text{mol } C_{12}H_{26}}$$

Chapter 6

Atomic Structure

Exercises

6.1 $0.9890 \times 12.000\ u = 11.868\ u$
$0.0110 \times 13.00335\ u = \underline{0.143\ u}$
$12.011\ u \Rightarrow 12.01\ u$

6.2 $0.9051 \times 19.99244\ u = 18.095\ u$
$0.0027 \times 20.99395\ u = 0.057\ u$
$0.0922 \times 21.99138\ u = \underline{2.028\ u}$
$20.180\ u \Rightarrow 20.18\ u$

6.3 $$\lambda = \frac{c}{\nu} = \frac{3.00 \times 10^{8}\ \text{m/s}}{9.76 \times 10^{13}\ \text{s}^{-1}} \times \frac{\text{nm}}{10^{-9}\ \text{m}} = 3.07 \times 10^{3}\ \text{nm}$$

6.4 $$\nu = \frac{c}{\lambda} = \frac{3.00 \times 10^{8}\ \text{m/s}}{1.07\ \text{mm}} \times \frac{\text{mm}}{10^{-3}\ \text{m}} = 2.80 \times 10^{11}\ \text{Hz}$$

6.5 $$E = h\nu = \frac{6.626 \times 10^{-34}\ \text{J s}}{\text{photon}} \times 2.89 \times 10^{10}\ \text{s}^{-1}$$

$E = 1.91 \times 10^{-23}$ J/photon

6.6 $$\lambda = \frac{hc}{E} = \frac{\dfrac{6.626 \times 10^{-34}\ \text{J s}}{\text{photon}} \times \dfrac{2.998 \times 10^{8}\ \text{m}}{\text{s}}}{\dfrac{1609\ \text{kJ}}{\text{mol photons}} \times \dfrac{10^{3}\ \text{J}}{\text{kJ}} \times \dfrac{\text{mol photons}}{6.022 \times 10^{23}\ \text{photons}}}$$

$$\lambda = 7.434 \times 10^{-8}\ \text{m} \times \frac{\text{nm}}{10^{-9}\ \text{m}} = 74.34\ \text{nm}$$

6.7 $$E_6 = \frac{-B}{n^2} = \frac{-2.179 \times 10^{-18}\ \text{J}}{6^2} = -6.053 \times 10^{-20}\ \text{J}$$

6.8 $$\Delta E = B \times \left(\frac{1}{n_i^2} - \frac{1}{n_f^2}\right) = 2.179 \times 10^{-18} \times \left(\frac{1}{2^2} - \frac{1}{4^2}\right)$$

$$= 2.179 \times 10^{-18} \times (0.1875) = 4.086 \times 10^{-19}\ \text{J}$$

6.9 $\Delta E = B \times \left(\frac{1}{n_i^2} - \frac{1}{n_f^2}\right) = 2.179 \times 10^{-18} \times \left(\frac{1}{5^2} - \frac{1}{2^2}\right)$

$$\Delta E = 2.179 \times 10^{-18} \times (-0.21) = -4.576 \times 10^{-19}\ \text{J}$$

$$\nu = \frac{\Delta E}{h} = \frac{4.576 \times 10^{-19}\ \text{J}}{6.626 \times 10^{-34}\ \text{J s}} = 6.906 \times 10^{14}\ \text{s}^{-1}$$

$$\lambda = \frac{c}{\nu} = \frac{2.998 \times 10^{8}\ \text{m/s}}{6.906 \times 10^{14}\ \text{s}^{-1}} = 4.341 \times 10^{-7}\ \text{m} \times \frac{\text{nm}}{10^{-9}\ \text{m}} = 434.1\ \text{nm}$$

6.10 $\lambda = \frac{h}{mu} = \frac{\frac{6.626 \times 10^{-34}\ \text{kg m}^2\ \text{s}}{\text{s}^2}}{1.67 \times 10^{-27}\ \text{kg} \times 3.79 \times 10^{3}\ \text{m/s}}$

$$\lambda = 1.05 \times 10^{-10}\ \text{m} \times \frac{\text{nm}}{10^{-9}\ \text{m}} = 0.105\ \text{nm}$$

6.11
a. not possible. For $l = 1$, m_l must be between +1 and -1.
b. All values are possible.
c. All values are possible.
d. not possible. For $l = 2$, m_l must be between +2 and -2.

6.12

a.	5 p $l = 1$	$m_l = -1, 0, 1$	3 orbitals
b.	$l = 0$	$m_l = 0$	1 orbital
	$l = 1$	$m_l = -1, 0, 1$	3 orbitals
	$l = 2$	$m_l = -2, -1, 0, 1, 2$	5 orbitals
	$l = 3$	$m_l = -3, -2, -1, 0, 1, 2, 3$	7 orbitals
	$n^2 = 4^2 = 16$ orbitals		16 orbitals
c.	$l = 3$	$m_l = -3, -2, -1, 0, 1, 2, 3$	f subshell consists of seven orbitals

Estimation Exercises

6.1 If the atomic weight is about 115 u and one isotope is 113 u, the other isotope must be at least ^{115}In.

6.2 $E = \frac{hc}{\lambda}$ $E\lambda = hc$ $E_1\lambda_1 = E_2\lambda_2$

$$\lambda_2 = \frac{E_1\lambda_1}{E_2} = \frac{\frac{189\ \text{kJ}}{\text{mol}}}{\frac{100\ \text{kJ}}{\text{mol}}} \times 632.8\ \text{nm} =$$

By estimation, $\lambda_2 = \frac{200}{100} \times 630$, $\lambda_2 \approx 1230$ nm or infrared.

Actual $\lambda_2 = 1196$ nm or infrared.

Conceptual Exercises

6.1 (c) requires energy to be emitted. Since the energy difference between two successive energy levels is smaller the higher the energy levels, (a) is larger than (d). Comparing (a) and (b) produces $\left(\frac{1}{1^2} - \frac{1}{2^2}\right)$ versus $\left(\frac{1}{3^2} - \frac{1}{\infty^2}\right)$ or 0.75 versus 0.11, so (a) represents the greatest amount of energy absorbed.

Review Questions

22. a. $3d$ b. $2s$ c. $4p$ d. $4f$

23. a. $n = 3$ $l = 0,\ l = 1,\ l = 2$ 3 sublevels

b. $n = 2$ $l = 0,\ l = 1$ 2 sublevels

c. $n = 4$ $l = 0,\ l = 1,\ l = 2,\ l = 3$ 4 sublevels

Problems

27. $1.044 \times 10^{-8}\ \text{kg/C} \times 1.602 \times 10^{-19}\ \text{C} = 1.672 \times 10^{-27}\ \text{kg}$

29. $6.4 \times 10^{-19}\ \text{C} = 4 \times 1.6 \times 10^{-19}\ \text{C}$
$3.2 \times 10^{-19}\ \text{C} = 2 \times 1.6 \times 10^{-19}\ \text{C}$
$4.8 \times 10^{-19}\ \text{C} = 3 \times 1.6 \times 10^{-19}\ \text{C}$

All values are integral multiples of 1.6×10^{-19} C, the charge on an electron.

31. a. $^{80}Br^-$ 80:1
b. $^{18}O^{2-}$ 9:1
c. $^{40}Ar^+$ 40:1
The masses used are mass numbers and not exact isotopic masses

33. Eu - 151 $0.478 \times 150.92 = 72.1$
Eu - 153 $0.522 \times 152.92 = \underline{79.8}$
151.9 u average atomic mass of Eu

35. Sr - 84 $.0056 \times 83.913 = 0.47$
Sr - 86 $.0986 \times 85.909 = 8.47$
Sr - 87 $.0700 \times 86.909 = 6.08$
Sr - 88 $.8258 \times 87.906 = \underline{72.59}$
87.61 u average atomic mass of Sr

37. $$4.4 \times 10^9\ \text{km} \times \frac{10^3\ \text{m}}{\text{km}} \times \frac{\text{s}}{3.00 \times 10^8\ \text{m}} = 1.5 \times 10^4\ \text{s}$$

39. $$\lambda = \frac{c}{\nu} = \frac{2.998 \times 10^8\ \text{m/s}}{992\ \text{k Hz} \times \frac{10^3\ \text{Hz}}{\text{k Hz}}} = 302\ \text{m}$$

41. $$\lambda = \frac{c}{\nu} = \frac{2.998 \times 10^8\ \text{m/s}}{9.74 \times 10^{18}\ \text{Hz}} = 3.08 \times 10^{-11}\ \text{m} \times \frac{\text{nm}}{10^{-9}\ \text{m}} = 0.0308\ \text{nm}$$

43. $\nu = \frac{c}{\lambda} = \frac{2.998 \times 10^8 \text{ m/s}}{539 \text{ nm} \times \frac{10^{-9}\text{ m}}{\text{nm}}} = 5.56 \times 10^{14} \text{ s}^{-1}$

45. $\lambda = \frac{c}{\nu} = \frac{2.998 \times 10^8 \text{ m/s}}{2.67 \text{ h}^{-1} \times \frac{\text{h}}{3600 \text{ s}}} = 4.04 \times 10^{11} \text{ m}$

$\lambda = \frac{c}{\nu} = \frac{2.998 \times 10^8 \text{ m/s}}{4.00 \text{ h}^{-1} \times \frac{\text{h}}{3600 \text{ s}}} = 2.70 \times 10^{11} \text{ m}$

47. $E = h\nu = 6.626 \times 10^{-34} \text{ J s} \times 3.73 \times 10^{14} \text{ s}^{-1} = 2.47 \times 10^{-19} \text{ J}$

49. $\nu = \frac{c}{\lambda} = \frac{2.998 \times 10^8 \text{ m/s}}{1.20 \text{ cm} \times \frac{10^{-2}\text{ m}}{\text{cm}}} = 2.498 \times 10^{10} \text{ s}^{-1}$

$E = h\nu = \frac{6.626 \times 10^{-34} \text{ J s}}{\text{photon}} \times 2.498 \times 10^{10} \text{ s}^{-1} = \frac{1.66 \times 10^{-23} \text{ J}}{\text{photon}}$

$\frac{1.66 \times 10^{-23} \text{ J}}{\text{photon}} \times \frac{6.022 \times 10^{23} \text{ photons}}{\text{mol}} \times \frac{\text{kJ}}{10^3 \text{ J}} = 0.0100 \text{ kJ /mol}$

51. $\nu = \frac{E}{h} = \frac{4.65 \times 10^{-19} \text{ J}}{6.626 \times 10^{-34} \text{ Js}} = 7.018 \times 10^{14} \text{ s}^{-1}$

$\lambda = \frac{c}{\nu} = \frac{2.998 \times 10^8 \text{ m/s}}{7.018 \times 10^{14} \text{ s}^{-1}} = 4.27 \times 10^{-7} \text{ m} \times \frac{\text{nm}}{10^{-9} \text{ m}} = 427 \text{ nm}$

53. $\frac{487 \text{ kJ}}{\text{mol}} \times \frac{\text{mol}}{6.022 \times 10^{23} \text{ photons}} \times \frac{10^3 \text{ J}}{\text{kJ}} = 8.087 \times 10^{-19} \text{ J/photon}$

$\nu = \frac{E}{h} = \frac{8.087 \times 10^{-19} \text{ J/photon}}{6.626 \times 10^{-34} \text{ J s/photon}} = 1.220 \times 10^{15} \text{ s}^{-1}$

$\lambda = \frac{c}{\nu} = \frac{2.998 \times 10^8 \text{ m/s}}{1.220 \times 10^{15} \text{ s}^{-1}} \times \frac{\text{nm}}{10^{-9} \text{ m}} = 246 \text{ nm};$ ultraviolet region

55. a. $E_i = \frac{-B}{n^2} = \frac{-2.179 \times 10^{-18}\ J}{3^2} = -2.421 \times 10^{-19}\ J$

$E_f = \frac{-B}{n^2} = \frac{-2.179 \times 10^{-18}\ J}{2^2} = -5.448 \times 10^{-19}\ J$

$\Delta E = E_f - E_i = -3.027 \times 10^{-19}\ J$

$\nu = \frac{\Delta E}{h} = \frac{3.027 \times 10^{-19}\ J}{6.626 \times 10^{-34}\ J\ s} = 4.568 \times 10^{14}\ s^{-1}$

b. $E_i = \frac{-B}{n^2} = \frac{-2.179 \times 10^{-18}\ J}{4^2} = -1.362 \times 10^{-19}\ J$

$E_f = \frac{-B}{n^2} = \frac{-2.179 \times 10^{-18}\ J}{1^2} = -2.179 \times 10^{-18}\ J$

$\Delta E = E_f - E_i = -2.043 \times 10^{-18}\ J$

$\nu = \frac{\Delta E}{h} = \frac{2.043 \times 10^{-18}\ J}{6.626 \times 10^{-34}\ J\ s} = 3.083 \times 10^{15}\ s^{-1}$

57. $E_1 = \frac{-2.179 \times 10^{-18}\ J}{1^2} = -2.179 \times 10^{-18}\ J$

$E_2 = \frac{-2.179 \times 10^{-18}\ J}{2^2} = -5.448 \times 10^{-19}\ J$

$E_3 = \frac{-2.179 \times 10^{-18}\ J}{3^2} = -2.421 \times 10^{-19}\ J$

$E_4 = \frac{-2.179 \times 10^{-18}\ J}{4^2} = -1.362 \times 10^{-19}\ J$

$E_5 = \frac{-2.179 \times 10^{-18}\ J}{5^2} = -8.716 \times 10^{-20}\ J$

$E_6 = \frac{-2.179 \times 10^{-18}\ J}{6^2} = -6.053 \times 10^{-20}\ J$

$E_7 = \frac{-2.179 \times 10^{-18}\ J}{7^2} = -4.447 \times 10^{-20}\ J$

$E_\infty = 0$

59. $\lambda = \frac{h}{mu} = \frac{6.626 \times 10^{-34}\ kg\ m^2\ s^{-1}}{1.67 \times 10^{-27}\ kg \times 2.55 \times 10^6\ m\ s^{-1}} = 1.56 \times 10^{-13}\ m \times \frac{nm}{10^{-9}\ m}$
$= 1.56 \times 10^{-4}$ nm.

61. $u = \frac{h}{m\lambda} = \frac{6.626 \times 10^{-34}\ kg\ m^2\ s^{-1}}{9.109 \times 10^{-31}\ kg \times 84.4\ nm \times \frac{10^{-9}\ m}{nm}} = 8.62 \times 10^3$ m/s

63. For f orbitals $l = 3$ $m = -3, -2, -1, 0, 1, 2, 3$ 7 4*f* orbitals

65. For *p* orbitals, $l = 1$ n must be 2 for $l = 1$.

67. If $n = 5$, $l = 0, 1, 2, 3, 4$.
If $l = 3$, $m_l = -3, -2, -1, 0, 1, 2, 3$.

69. a. permissible
b. If $n = 3$, l cannot be 3; l cannot be greater than n-1.
c. permissible
d. $n = 0$ is not permissible.

71. a. For 3*s*, $n = 3, l = 0, m_l = 0$
b. For 5*f*, $n = 5, l = 3, m_l = -3, -2, -1, 0, 1, 2, 3$
c. 3*p*

73. 174.967 = 174.941 X + (1.00 - X) 175.943
174.967 = 174.941 X+ 175.943 - 175.943 X
1.002 X = 0.976
X = 0.974 for ^{175}Lu
0.026 for ^{176}Lu

75. $T = \frac{1}{\nu}$
$\frac{1s}{60\ cycles} = 1.7 \times 10^{-2}$ s/cycle
Period and frequency are inversely related.

77. Each photon that has more energy than the threshold energy will dislodge an electron. The energies of two photons do not add together to reach the threshold energy. Each photon interacts with the metal separately.

79. $\Delta E = B\left(\frac{1}{n_i^2} - \frac{1}{n_f^2}\right)$

If level 1 –> 2 $\Delta E = B\left(\frac{1}{1} - \frac{1}{4}\right) = \frac{3}{4}B$,

then level 1 –> 3 c. $\Delta E = B\left(1 - \frac{1}{9}\right) = \Delta E = \frac{8}{9}B = \frac{32}{27} \times \frac{3}{4}B$

81. $\Delta E = E_\infty - E_1 = 0 - \dfrac{-2.179 \times 10^{-18}\text{ J}}{1^2} = 2.179 \times 10^{-18}\text{ J}$

$\lambda = \dfrac{hc}{E} = \dfrac{6.626 \times 10^{-34}\text{ J s} \times 2.998 \times 10^{8}\text{ m/s}}{2.179 \times 10^{-18}\text{ J}} \times \dfrac{\text{nm}}{10^{-9}\text{ m}} = 91.16\text{ nm}$ minimum

$\Delta E = E_2 - E_1 = \dfrac{-2.179 \times 10^{-18}\text{ J}}{2^2} - \dfrac{-2.179 \times 10^{-18}\text{ J}}{1^2} = 1.634 \times 10^{-18}\text{ J}$

$\lambda = \dfrac{hc}{E} = \dfrac{6.626 \times 10^{-34}\text{ J s} \times 2.998 \times 10^{8}\text{ m/s}}{1.634 \times 10^{-18}\text{ J}} \times \dfrac{\text{nm}}{10^{-9}\text{ m}} = 121.6\text{ nm}$ maximum

83. $E = \dfrac{hc}{\lambda} = \dfrac{6.626 \times 10^{-34}\text{ J s} \times 2.998 \times 10^{8}\text{ m/s}}{486.1\text{ nm} \times 10^{-9}\text{ m/nm}} = 4.087 \times 10^{-19}\text{ J}$

$\Delta E = \dfrac{-B}{n^2} - \dfrac{-B}{m^2} = -B\left(\dfrac{1}{n^2} - \dfrac{1}{m^2}\right)$

$4.087 \times 10^{-19}\text{ J} = -2.179 \times 10^{-18}\text{ J} \times \left(\dfrac{1}{n^2} - \dfrac{1}{m^2}\right)$

$\dfrac{1}{n^2} - \dfrac{1}{m^2} = 0.1876$

$\dfrac{1}{1^2} = 1 \quad \dfrac{1}{2^2} = 0.2500 \quad \dfrac{1}{3^2} = 0.1111 \quad \dfrac{1}{4^2} = 0.0625 \quad \dfrac{1}{5^2} = 0.0400$

$\dfrac{1}{2^2} - \dfrac{1}{4^2} = 0.1875$ Thus, the transition is $n = 4 \longrightarrow n = 2$

85. $\Delta E = E_2 - E_1 = \dfrac{-2.179 \times 10^{-18}\text{ J}}{2^2} - \dfrac{-2.179 \times 10^{-18}\text{ J}}{1^2} = 1.634 \times 10^{-18}\text{ J}$

$\Delta E = E_3 - E_2 = \dfrac{-2.179 \times 10^{-18}\text{ J}}{3^2} - \dfrac{-2.179 \times 10^{-18}\text{ J}}{2^2} = 3.026 \times 10^{-19}\text{ J}$

total = 1.937 x 10-18 J

$\Delta E = E_3 - E_1 = \dfrac{-2.179 \times 10^{-18}\text{ J}}{3^2} - \dfrac{-2.179 \times 10^{-18}\text{ J}}{1^2} = 1.937 \times 10^{-18}\text{ J}$

The sum of the energies of the transitions: $n = 2 \longrightarrow n = 1$ ($1.634\ 10^{-18}$ J) and $n = 3 \longrightarrow n = 2$ (3.026×10^{-19} J) is the same as the energy of the transition: $n = 3 \longrightarrow n = 1$ (1.937×10^{-18} J).

87. $\lambda = \frac{h}{mu} = \frac{6.626 \times 10^{-34}\ \text{J s}}{25.0\ \text{g} \times \frac{\text{kg}}{10^3\ \text{g}} \times 110\ \text{m/s}} = 2.41 \times 10^{-34}\ \text{m}$

89. $\lambda = \frac{h}{mu}$ The mass times the velocity must be equal for the same wavelength to be produced by both the electron and the proton. Because the electron has the smaller mass, it must have the greater velocity.

$$\frac{4 \times 10^{-21}\ \text{watt}}{5.566 \times 10^{-24}\ \text{J /photon}} \times \frac{\text{J}}{\text{watt s}} = 7 \times 10^{2}\ \text{photon/s}$$

91. $q = m\text{sp.ht.}\Delta T$

$q = 345\ \text{g} \times \frac{4.18\ \text{J}}{\text{g °C}} \times (99.8 - 26.5)\text{°C} = 1.057 \times 10^{5}\ \text{J}$

$E = h\nu = 6.626 \times 10^{-34}\ \text{J s} \times 2.45 \times 10^{9}\ \text{s}^{-1} = 1.623 \times 10^{-24}\ \text{J/photon}$

$$\frac{1.057 \times 10^{5}\ \text{J}}{1.623 \times 10^{-24}\ \text{J/photon}} \times \frac{1\ \text{mol photons}}{6.022 \times 10^{23}\ \text{photon}} =$$

1.08×10^{5} mol photons

93. Bohr meant that Einstein should accept some things as unknowable, that he should not assume that all natural phenomena were meant to be predictable.

Chapter 7

Electron Configurations and the Periodic Table

Exercises

7.1 a. This notation indicates two electrons in the 1*s* subshell, two in the 2*s* subshell, six in the 2*p* subshell, two in the 3*s* subshell, six in the 3*p* subshell, ten in the 3*d* subshell, and two in the 4*s* subshell. This is the electron configuration of zinc (Z = 30).

b. This orbital diagram has two electrons in the 1*s* subshell, two electrons in the 2*s* subshell, two electrons in each of the 2*p* orbitals, two electrons in the 3*s* subshell, and one electron in each of two 3*p* orbitals. Each pair of arrows represents two electrons with opposite spins. The electrons in the 3*p* orbitals have parallel spins. This element is silicon.

7.2 P $1s^22s^22p^63s^23p^3$

[Ne] $3s^23p^3$

3*s* 3*p*

[Ne] [↑↓] [↑][↑][↑]

7.3 Ga $4s^24p^1$

Te $5s^25p^4$

7.4 a. Ge $1s^22s^22p^63s^23p^63d^{10}4s^24p^2$

Ge [Ar]$3d^{10}4s^24p^2$

b. Zn $1s^22s^22p^63s^23p^63d^{10}4s^2$

Zn [Ar]$3d^{10}4s^2$

c. Ti $1s^22s^22p^63s^23p^63d^24s^2$

Ti [Ar]$3d^24s^2$

d. I $1s^22s^22p^63s^23p^63d^{10}4s^24p^64d^{10}5s^25p^5$

I [Kr]$4d^{10}5s^25p^5$

7.5 a. K [Ar]$4s^1$ paramagnetic

b. Hg [Xe]$4f^{14}5d^{10}6s^2$ diamagnetic

c. Ba [Xe]$6s^2$ diamagnetic

d. Ga [Ar]$3d^{10}4s^24p^1$ paramagnetic

3*s* 3*p*

e. S [Ne] [↑↓] [↑↓][↑][↑] paramagnetic

6*p*

f. Pb [Xe]$4f^{14}5d^{10}6s^2$ [↑][↑][ ] paramagnetic

7.6 a. F < N < Be
b. Be < Ca < Ba
c. F < Cl < S
d. Mg < Ca < K

7.7 All are isoelectronic, so the more protons, the smaller the size.
$Y^{3+} < Sr^{2+} < Rb^{+} < Br^{-} < Se^{2-}$

7.8 a. Be < N < F
b. Ba < Ca < Be
c. S < P< F
d. K < Ca < Mg

7.9 a. O is more nonmetallic.
b. S is more nonmetallic.
c. F is more nonmetallic.

Estimation Exercise
7.1 Because the Se atom is larger than the S atom, the value must be less than 450 kJ/mol but a positive number. The value must be 400 kJ/mol.

Conceptual Exercise
7.1

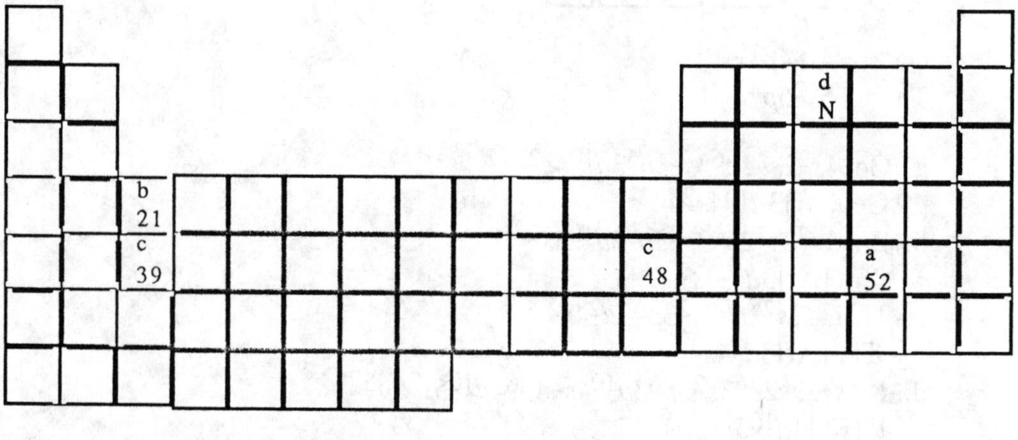

c. Cadmium is the highest; yttrium is the lowest.

Review Questions
3. The orbital energies in (b) and (c) are identical for the H atom; those in (c) are also identical for other atoms.

9. 3A elements $ns^2\, np^1$
5A elements $ns^2\, np_x^1\, np_y^1\, np_z^1$

10. a. *ns*; b. *np* c. (*n*-1)*d* d. (*n*-2)*f*.

12. a. 4 b. 8 c. 7 d. 3 e. 2

15. a. period 2nd group 8A
b. 3rd 4A
c. 3rd 1A
d. 2nd 2A
e. 2nd 5A
f. 3rd 3A

16. 4A ns^2np^2
6A ns^2np^4

18. 2 electrons in an orbital.
10 electrons in the *d* subshell.

19. a. 3*s* b. 3*s* c. 2*p* d. 3*p*

21. a. Si - $1s^22s^22p^6$ $3s^23p^2$
core valence

b. Rb - $1s^22s^22p^63s^23p^63d^{10}4s^24p^6$ $5s^1$
core valence

c. Br - $1s^22s^22p^63s^23p^63d^{10}$ $4s^24p^5$
core valence

23. a. diamagnetic b. diamagnetic
c. paramagnetic d. paramagnetic
e. diamagnetic f. paramagnetic

26. a. 5
b. 32 ($4s^24p^64d^{10}4f^{14}$)
c. 5 (nitrogen, phosphorus, arsenic, antimony, bismuth)
d. 2
e. 10 (yttrium, zirconium, niobium, molybdenium, technetium, ruthenium, rhodium, palladium, silver, cadmium)

27. a. 3 b. 1 c. 0 d. 14 e. 1 f. 5
g. 32 (cesium, barium, thallium, lead, bismuth, polonium, astatine, radon, lanthanum through mercury and the lanthanides.)

33. a. Cl < S b. Al < Mg c. As < Ge d. Ca < K

37. a. Sr < Ca < Mg b. S < P < Cl
c. Sn < Ge < As d. Se < Br < Cl

Problems

43. a. Allowed
b. Not allowed. There are three electrons in one orbital and the 2*s* electrons have the same spin.
c. Not allowed. There are three electrons in one of the 2*p* orbitals.
d. Not allowed. The unpaired electrons are not all of the same spin.
e. Not allowed. There should be one electron in each 3*p* orbital.
f. Allowed

45. a. The 3*s* subshell fills before any electron goes into the 3*p* subshell.
b. The 4*s* subshell fills before 3*d* subshell begins to fill.
c. The 3*d* subshell fills completely before 4*p* subshell begins to fill.

47. a. The subshell that fills after [Ar] is 4*s*. 2*d* does not exist.
b. The subshell that fills after [Ar] $4s^2$ is 3*d*. 3*f* does not exist.
c. The subshell that fills after [Kr] $4d^{10}5s^2$ is 5*p*. 4*f* does not fill until after 6*s*.

49.

a.	Ar	$1s^22s^22p^63s^23p^6$
b.	H	$1s^1$
c.	Ne	$1s^22s^22p^6$
d.	Be	$1s^22s^2$
e.	K	$1s^22s^22p^63s^23p^64s^1$
f.	P	$1s^22s^22p^63s^23p^3$
g.	Ca	$1s^22s^22p^63s^23p^64s^2$
h.	Mg	$1s^22s^22p^63s^2$
i.	Br	$1s^22s^22p^63s^23p^63d^{10}4s^24p^5$

51.

a.	Ga	[Ar] $3d^{10}4s^24p^1$
b.	Te	[Kr] $4d^{10}5s^25p^4$
c.	I	[Kr] $4d^{10}5s^25p^5$
d.	Cs	[Xe] $6s^1$
e.	Sb	[Kr] $4d^{10}5s^25p^3$
f.	Sr	[Kr] $5s^2$

53.

a. N — 1s [↑↓] 2s [↑↓] 2p [↑][↑][↑]

b. B — 1s [↑↓] 2s [↑↓] 2p [↑][ ][ ]

c. Si — 1s [↑↓] 2s [↑↓] 2p [↑↓][↑↓][↑↓] 3s [↑↓] 3p [↑][↑][ ]

d. Ca — 1s [↑↓] 2s [↑↓] 2p [↑↓][↑↓][↑↓] 3s [↑↓] 3p [↑↓][↑↓][↑↓] 4s [↑↓]

e. Cl — 1s [↑↓] 2s [↑↓] 2p [↑↓][↑↓][↑↓] 3s [↑↓] 3p [↑↓][↑↓][↑]

f. Sc — 1s [↑↓] 2s [↑↓] 2p [↑↓][↑↓][↑↓] 3s [↑↓] 3p [↑↓][↑↓][↑↓] 3d [↑][ ][ ][ ][ ] 4s [↑↓]

55. Hg [Xe] $4f^{14}5d^{10}6s^2$ Hg is in the 6th period and Group 2B, and it is the last 5*d* transition element.

[Xe] 4f [↑↓][↑↓][↑↓][↑↓][↑↓][↑↓][↑↓] 5d [↑↓][↑↓][↑↓][↑↓][↑↓] 6s [↑↓]

57. a. Ca < Sr < Rb
Strontium has one more shell of electrons than calcium. It has one more electron and proton than rubidium, so it is smaller. Atomic radius increases from top to bottom in a group of elements and from right to left in a period of elements.

b. C < Si < Al
Silicon has one more shell of electrons than carbon, so is larger. Silicon has one more electron and proton than aluminum, so is smaller. Atomic radius increases from top to bottom in a group of elements and from right to left in a period of elements.

59. a. K < Ca < Mg
A potassium atom is larger than a calcium atom, which is larger than a magnesium atom.

b. Te < I < Br
A tellurium atom is larger than an iodine atom, which is larger than a bromine atom..

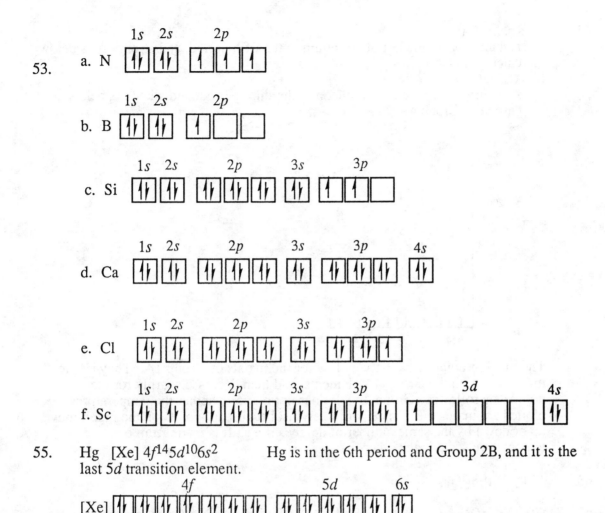

61. a. Rb, Sr, Ca
Rubidium is to the left of strontium in the fifth period, and strontium is below calcium in Group 2A.
b. Al, Si, C
Aluminum is to the left of silicon in the third period, and silicon is below carbon in Group 4A.

63.

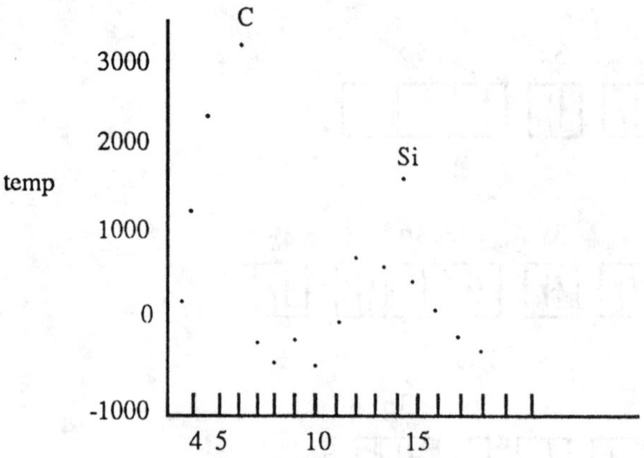

The melting points are relatively low for the metals of Group 1A. They rise to a maximum with the Group 4A elements, and then decrease sharply for the nonmetals following Group 4A. It does not matter whether the temperatures plotted are in degrees Celsius or in kelvins, because all points on the graph move up or down by the same number of degrees if a shift is made from one temperature scale to the other.

65. 114 [Rn] $5f^{14}6d^{10}7s^27p^2$

67. $I_1(Cs) < I_1(B) < I_2(Sr) < I_2(In) < I_2(Xe) < I_3(Ca)$. A Cs atom is larger than a B atom and has a lower I_1. Second ionization energies of atoms are larger than their first ionization energies. $I_2(Sr)$ should be larger than $I_1(B)$, but it should be smaller than the other I_2 values, which increase from left to right in the fifth period. $I_3(Ca)$ should be greatest of all because it represents removing an electron from the noble-gas electron configuration of Ca^{2+}.

69. $Cl\ (g) + e^- \longrightarrow Cl^-\ (g)$ ΔH = electron affinity = -349 kJ.

For the overall process, $\Delta H = \left(\frac{1}{2} \times 242.8 \text{ kJ}\right) - 349 \text{ kJ} = -228 \text{ kJ/mol } Cl^-$. The overall process is exothermic.

71.

$Na^+(g) + e^- \longrightarrow Na\ (g)$	- 496 kJ
$Cl^-\ (g) \longrightarrow e^- + Cl\ (g)$	- (- 349) kJ
$Na^+\ (g) + Cl^-\ (g) \longrightarrow Na\ (g) + Cl\ (g)$	- 147 kJ exothermic

73. $$\frac{1 \text{eV}}{\text{atom}} \times \frac{6.0221 \times 10^{23} \text{ atoms}}{\text{mol}} \times \frac{1.6022 \times 10^{-19} \text{ C}}{\text{electron}} \times \frac{1 \text{ J}}{1 \text{V}\cdot\text{C}} \times \frac{1 \text{ kJ}}{10^3 \text{ J}} = 96.49 \text{ kJ/mol}$$

Chapter 8

The s-Block Elements

Exercises

8.1 $Zn\ (s) + 2\ H^+\ (aq) \rightarrow Zn^{2+}\ (aq) + H_2\ (g)$

$$1.00 \text{ kg Zn} \times \frac{10^3 \text{ g}}{\text{kg}} \times \frac{\text{mol Zn}}{65.39 \text{ g Zn}} \times \frac{\text{mol } H_2}{\text{mol Zn}} \times \frac{2.016 \text{ g } H_2}{\text{mol } H_2} = 30.8 \text{ g } H_2$$

Zinc produces more H_2 than is produced from spent uranium fuel (25.4 g H_2), but producing H_2 from spent fuels is getting extra value from spent fuel before it has to be stored anyway.

8.2 a. $2\ NaCl\ (l) \xrightarrow{\text{electrolysis}} 2\ Na\ (s) + Cl_2\ (g)$

$2\ Na\ (s) + H_2\ (g) \xrightarrow{\Delta} 2\ NaH\ (s)$

b. $2\ NaCl\ (aq) + 2\ H_2O\ (l) \xrightarrow{\text{electrolysis}} 2\ NaOH\ (aq) + Cl_2\ (g) + H_2\ (g)$

$2\ NaOH\ (aq) + Cl_2\ (g) \rightarrow NaCl\ (aq) + NaOCl\ (aq) + H_2O\ (l)$

8.3 $3\ BaO\ (s) + 2\ Al\ (l) \xrightarrow{1800°C} 3\ Ba\ (g) + Al_2O_3\ (s)$

Estimation Exercises

8.1 a. $Zn\ (s) + 2\ H^+ \rightarrow Zn^{2+} + H_2\ (g)$

$$\frac{1 \text{ mole } H_2}{1 \text{ mole Zn}} = \frac{2 \text{ g } H_2}{65 \text{ g Zn}}$$

b. $2\ Na\ (s) + 2\ H_2O\ (l) \rightarrow 2\ NaOH\ (aq) + H_2\ (g)$

$$\frac{1 \text{ mole } H_2}{2 \text{ mole Na}} = \frac{2 \text{ g } H_2}{2 \times 23 \text{ g Na}}$$

c. $CaH_2\ (s) + 2\ H_2O\ (l) \rightarrow Ca(OH)_2\ (aq) + 2\ H_2\ (g)$

$$\frac{2 \text{ mole } H_2}{\text{mole } CaH_2} = \frac{2 \times 2 \text{ g } H_2}{42 \text{ g} CaH_2}$$

The comparison is very simple -

Zn	Na	CaH_2
$\frac{2}{65}$	$\frac{2}{23 \times 2}$	$\frac{2 \times 2}{42}$

CaH_2 obviously produces the most H_2 from a given mass of the reactant.

8.2 The largest mass percent of water would be in the hydrate with the smallest formula mass. Cl has a lower atomic weight than Br, and Mg has a lower atomic weight than Ca or Sr, so $MgCl_2 \cdot 6H_2O$ has the largest mass percent of water.

Conceptual Esercise

8.1 Let X be the mass of Mg that reacts to form MgO and $0.267 - X$, the mass of Mg that reacts to form Mg_3N_2.

$$?g\ MgO = X \times \frac{1\ mol\ Mg}{24.31\ g\ Mg} \times \frac{2\ mol\ MgO}{2\ mol\ Mg} \times \frac{40.30\ g\ MgO}{1\ mol\ MgO} = 1.658\ X$$

$$?g\ Mg_3N_2 = (0.267\text{-}X) \times \frac{1\ mol\ Mg}{24.31\ g\ Mg} \times \frac{1\ mol\ Mg_3N_2}{3\ mol\ Mg} \times \frac{101.0\ g\ Mg_3N_2}{1\ mol\ Mg_3N_2} =$$

$1.385 \times (0.267 - X) = 0.370 - 1.385\ X$

The total mass of product is 0.420g. That is,

$1.658\ X + (0.370 - 1.385\ X) = 0.420$

$0.273\ X = 0.050$

$X = 0.18$

Mass of MgO = $1.658\ X = 1.658 \times 0.18 = 0.30$ g

$$\text{Mass \% MgO} = \frac{0.30\ g\ MgO \times 100\%}{0.420\ g\ products} = 71\%\ MgO.$$

An alternate method of working the problem produces a slightly different answer because two numbers that are close are being subtracted.

Let M be the mass fraction of MgO formed and $(1.00 - M)$, the mass fraction of Mg_3N_2.

$0.443\ g \times M + 0.370\ g \times (1.00\text{-}M) = 0.420\ g$

$0.073 \times M + 0.370 = 0.420$

$0.073 \times M = 0.050$

$M = 0.68$

Mass % MgO = 68% MgO

Review Questions

1. oxygen, hydrogen, hydrogen

3. Sodium is the most abundant alkali metal and calcium is the most abundant alkaline earth metal.

5. The alkali metal ions essential to living organisms are sodium and potassium. The essential alkaline earth metal ions are magnesium and calcium.

6. In Group 1A, lithium, sodium, potassium, rubidium, cesium and francium all react with cold water. In Group 2A, the metals reacting with cold water are calcium, strontium, barium, and radium.

15. a. Magnesium metal is produced in the Dow process.
b. Sodium hydrogen carbonate is produced in the Solvay process, and is usually converted to sodium carbonate.

16. a. calcium carbonate, $CaCO_3$
b. calcium oxide, CaO
c. calcium hydroxide, $Ca(OH)_2$
d. calcium sulfate dihydrate, $CaSO_4 \cdot 2H_2O$
e. sodium sulfate decahydrate, $Na_2SO_4 \cdot 10H_2O$

17. a. Portland cement: limestone, clay, and sand
    b. Soda-lime glass: limestone, sand. and sodium carbonate
    c. Mortar: slaked lime, sand, and water
    d. Plaster of Paris: gypsum

Problems

27. a. lithium hydride
    b. MgI_2
    c. $Ba(OH)_2 \cdot 8H_2O$
    d. calcium hydrogen carbonate

29. a. $Ca(s) + 2\,HCl(aq) \longrightarrow H_2(g) + CaCl_2(aq)$
    b. $CaO(s) + 2\,HCl(aq) \longrightarrow CaCl_2(aq) + H_2O(l)$
    c. $2\,NaF(s) + H_2SO_4(\text{conc. aq}) \xrightarrow{\Delta} Na_2SO_4(s) + 2\,HF(g)$

31. a. $2\,Li(s) + Cl_2(g) \longrightarrow 2\,LiCl(s)$
    b. $2\,K(s) + 2\,H_2O \longrightarrow 2\,KOH(aq) + H_2(g)$
    c. $2\,Cs(s) + Br_2(l) \longrightarrow 2\,CsBr(s)$
    d. $K(s) + O_2(g) \longrightarrow KO_2(s)$

33. a. $MgCO_3(s) \xrightarrow{\Delta} MgO(s) + CO_2(g)$
    b. $CaCl_2(l) \xrightarrow{\text{electrolysis}} Ca(s) + Cl_2(g)$
    c. $Ca(s) + 2\,HCl(aq) \longrightarrow CaCl_2(aq) + H_2(g)$
    d. $Ca(OH)_2(s) + H_2SO_4(aq) \longrightarrow CaSO_4(s) + 2\,H_2O(l)$

35. a. $Mg(OH)_2(s) + 2\,HCl(aq) \longrightarrow MgCl_2(aq) + 2H_2O(l)$
    b. $CO_2(g) + 2\,KOH(aq) \longrightarrow K_2CO_3(aq) + H_2O(l)$
    c. $2\,KCl(s) + H_2SO_4(\text{conc. aq}) \longrightarrow K_2SO_4(s) + 2\,HCl(g)$

37. a. $2\,NaCl(s) + H_2SO_4(\text{conc. aq}) \xrightarrow{\Delta} Na_2SO_4(s) + 2\,HCl(g)$
    b. $Mg(HCO_3)_2(aq) \xrightarrow{\Delta} MgCO_3(s) + H_2O(l) + CO_2(g)$
       $MgCO_3(s) \xrightarrow{\Delta} MgO(s) + CO_2(g)$

39. a. $2\,NH_4Cl(aq) + Ca(OH)_2(aq) \longrightarrow 2\,NH_3(g) + CaCl_2(aq) + 2\,H_2O(l)$
    b. $CaCO_3(s) \xrightarrow{\Delta} CaO(s) + CO_2(g)$ followed by:
       $CaO(s) + H_2O(l) \longrightarrow Ca(OH)_2(aq)$ (an alkaline solution)

41. $CH_3CH = CH_2 + H_2 \longrightarrow C_3H_8$ $\qquad V = \dfrac{nRT}{P}$

$$175\text{ g } C_3H_6 \times \frac{\text{mol } C_3H_6}{42.08\text{ g}} \times \frac{\text{mol } H_2}{\text{mol } C_3H_6} \times \frac{\dfrac{0.08206\text{ L atm}}{\text{K mol}} \times 295\text{ K}}{15.5\text{ atm}} = 6.50\text{ L } H_2$$

$$175\text{ g } C_3H_6 \times \frac{\text{mol } C_3H_6}{42.08\text{ g}} \times \frac{\text{mol } C_3H_8}{\text{mol } C_3H_6} \times \frac{44.10\text{ g}}{\text{mol } C_3H_8} = 183\text{ g } C_3H_8$$

43. $$1.00 \text{ kg Mg} \times \frac{10^3 \text{ g}}{\text{kg}} \times \frac{\text{ton seawater}}{1270 \text{ g Mg}} \times \frac{2000 \text{ lb}}{\text{ton}} \times \frac{454 \text{ g}}{\text{lb}} \times \frac{\text{mL}}{1.03 \text{ g}} \times \frac{10^{-3} \text{ L}}{\text{mL}}$$

$$= 694 \text{ L seawater}$$

The actual volume required is greater than 694 L because not every step in the overall process has a 100% yield.

45. $CaCO_3 \cdot MgCO_3 \longrightarrow CaO \cdot MgO + 2\ CO_2\ (g)$

$$1.00 \times 10^3 \text{ kg} \times \frac{10^3 \text{ g}}{\text{kg}} \times \frac{\text{mol } CaCO_3 \cdot MgCO_3}{184.4 \text{ g}} \times \frac{2 \text{ mol } CO_2}{\text{mol}} \times \frac{0.08206 \text{ atm}}{\text{K mol}} \times$$

$$\frac{295 \text{ K}}{748 \text{ mmHg} \times \frac{\text{atm}}{760 \text{ mmHg}}} \times \frac{1 \text{ m}^3}{10^3 \text{ L}} = 267 \text{ m}^3\ CO_2$$

47. Permanent hard water is not softened by addition of NH_3. No carbonate ion is formed to yield precipitates of MCO_3, and the concentration of OH^- in the weakly basic solution is not high enough to cause the precipitation of insoluble $M(OH)_2$.

49. $Ca(OH)_2 + 2\ HCO_3^- + Ca^{2+} \longrightarrow 2\ CaCO_3\ (s) + 2\ H_2O$
Note that each formula unit of $Ca(OH)_2$ is capable of converting two HCO_3^- ions to CO_3^{2-} ions. Two formula units of $CaCO_3$ precipitate, with half of the required Ca^{2+} coming from the $Ca(OH)_2$ and half from the hard water.

$$1.50 \times 10^7 \text{ L water} \times \frac{126 \text{ g } HCO_3^-}{10^3 \text{ L water}} \times \frac{\text{mol } HCO_3^-}{61.02\text{g}} \times \frac{\text{mol } Ca(OH)_2}{2 \text{ mol } HCO_3^-} \times \frac{74.10 \text{ g}}{\text{mol } Ca(OH)_2}$$

$$\times \frac{\text{kg}}{10^3 \text{ g}} = 1.15 \times 10^3 \text{ kg } Ca(OH)_2$$

51. Some of the soap must be precipitated as a scum by hard water cations. Only after this occurs can additional soap be effective as a cleaning agent.

53. a. Ca^{2+} + resin - 2 H $\longrightarrow$ resin - Ca + 2 H^+
HCO_3^- + resin - OH $\longrightarrow$ resin - HCO_3 + OH^-
$H^+ + OH^- \longrightarrow H_2O$

b. Na^+ + resin - H $\longrightarrow$ resin - Na + H^+
Cl^- + resin - OH $\longrightarrow$ resin - Cl + OH^-
$H^+ + OH^- \longrightarrow H_2O$

c. Na^+ + resin - H $\longrightarrow$ resin - Na + H^+
$H^+ + OH^- \longrightarrow H_2O$

55. a. $4\ Li\ (s) + O_2\ (g) \longrightarrow 2\ Li_2O(s)$
b. $6\ Li\ (s) + N_2\ (g) \longrightarrow 2\ Li_3N\ (s)$
c. $Li_2CO_3\ (s) \longrightarrow Li_2O(s) + CO_2\ (g)$

57.

$$
\begin{array}{ccccc}
 & & CaCO_3 & & \\
 & & \uparrow\ CO_2\,(g) & & \\
CaHPO_4 & \xleftarrow{H_3PO_4\,(aq)} & Ca(OH)_2 & \xrightarrow{HCl\,(aq)} & CaCl_2 \\
 & & \downarrow\ H_2SO_4\,(aq) & & \\
 & & CaSO_4 & &
\end{array}
$$

59. a. $Ca^{2+} + Na_2CO_3 \longrightarrow CaCO_3\,(s) + 2Na^+$

b. $162\text{ L water}\,\dfrac{117\text{ g }SO_4^{2-}}{10^3\text{ L water}} \times \dfrac{\text{mol }SO_4^{2-}}{96.06\text{ g}} \times \dfrac{\text{mol }Ca^{2+}}{\text{mol }SO_4^{2-}} \times \dfrac{\text{mol }Na_2CO_3}{\text{mol }Ca^{2+}} \times$

$\dfrac{105.99\text{ g}}{\text{mol}} = 20.9\text{ g }Na_2CO_3$

61. $\dfrac{185\text{ g }Ca^{2+}}{10^6\text{ g solution}} \times \dfrac{10^6\text{ g}}{1000\text{ L}} \times \dfrac{\text{mol }Ca^{2+}}{40.08\text{ g }Ca^{2+}} \times \dfrac{2\text{ mol }Na^+}{\text{mol }Ca^{2+}} = 9.23 \times 10^{-3}\text{ M }Na^+$

63. a. $CaO + 2\,HCl \longrightarrow CaCl_2 + H_2O$

$985\text{ L} \times 1.38\text{ M} \times \dfrac{\text{mol CaO}}{2\text{ mol HCl}} \times \dfrac{56.08\text{ g CaO}}{\text{mol CaO}} \times \dfrac{\text{kg}}{10^3\text{ g}} = 38.1\text{ kg CaO}$

b. $NaOH + HCl \longrightarrow NaCl + H_2O$

$985\text{ L} \times 1.38\text{ M} \times \dfrac{\text{mol NaOH}}{\text{mol HCl}} \times \dfrac{40.00\text{ g NaOH}}{\text{mol NaOH}} \times \dfrac{\text{kg}}{10^3\text{ g}} = 54.4\text{ kg NaOH}$

65. $CaO\,(s) + CO_2\,(g) \longrightarrow CaCO_3\,(s)$ Carbon dioxide reacts with the CaO.
or $CaO\,(s) + H_2O\,(g) \longrightarrow Ca(OH)_2\,(s)$ followed by:
$Ca(OH)_2\,(s) + CO_2\,(g) \rightarrow CaCO_3\,(s) + H_2O\,(g)$ Water vapor reacts with the CaO.

67. a. $H_2\,(l) + \frac{1}{2}O_2\,(g) \longrightarrow H_2O$ $\quad \Delta H^\circ = -285.8\text{ kJ}$

In estimation exercise 5.2, ΔH combustion was estimated.

$C_{12}H_{26} + \frac{37}{2}O_2 \longrightarrow 12\,CO_2 + 13\,H_2O \quad \Delta H^\circ = -7900\text{ kJ}$

In estimation exercise 5.1, ΔH combustion was estimated.

H_2 - $\quad \dfrac{-285.8\text{ kJ}}{\text{mol}} \times \dfrac{\text{mol}}{2.016\text{ g}} = -141.8\,\dfrac{\text{kJ}}{\text{g}}$

$C_{12}H_{26}$ - $\quad \dfrac{-7900\text{ kJ}}{\text{mol}} \times \dfrac{\text{mol}}{170.3\text{ g}} = \dfrac{-46\text{ kJ}}{\text{g }C_{12}H_{26}}$

b. H_2 - $\quad \dfrac{-141.8\text{ kJ}}{\text{g}} \times \dfrac{0.0708\text{ g}}{\text{mL}} = \dfrac{-10.0\text{ kJ}}{\text{mL}}$

$C_{12}H_{26}$ - $\quad \dfrac{-46\text{ kJ}}{\text{g}} \times \dfrac{0.749\text{ g}}{\text{mL}} = \dfrac{-34\text{ kJ}}{\text{mL}}$

Chapter 9

Chemical Bonds

Exercises

9.1   a. $:\overset{\cdot}{\underset{\cdot\cdot}{Ar}}:$ b. $\cdot Ca \cdot$ c. $\cdot \overset{\cdot}{\underset{\cdot\cdot}{Br}}:$ d. $:\overset{\cdot}{As}\cdot$ e. $K\cdot$ f. $\cdot \overset{\cdot}{\underset{\cdot\cdot}{Se}}\cdot$

9.2

$$\cdot \overset{\cdot}{Al} \cdot \quad \cdot \overset{\cdot\cdot}{\underset{}{O}} \cdot$$
$$\cdot \overset{\cdot}{\underset{}{O}} \cdot \quad \longrightarrow \quad [Al]^{3+} \quad \left[:\overset{\cdot\cdot}{\underset{\cdot\cdot}{O}}: \right]^{2-}$$
$$\cdot \overset{\cdot}{Al} \cdot \quad \cdot \overset{\cdot\cdot}{\underset{}{O}} \cdot \qquad [Al]^{3+} \quad \left[:\overset{\cdot\cdot}{\underset{\cdot\cdot}{O}}: \right]^{2-}$$
$$\left[:\overset{\cdot\cdot}{\underset{\cdot\cdot}{O}}: \right]^{2-}$$

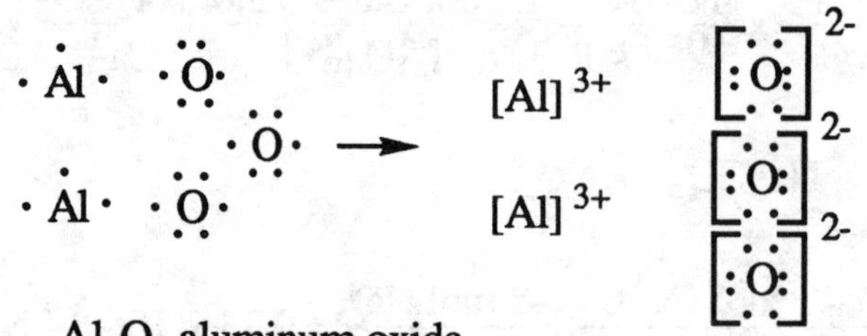

Al_2O_3 aluminum oxide

9.3 $Li^+ (g) + Cl^- (g) \rightarrow LiCl (s)$ $\Delta H_{LE} = ?$
$Li^+ (g) + e^- \rightarrow Li (g)$ $-$ (520 kJ)
$Li (g) \rightarrow Li (s)$ $-$ (161 kJ)
$Li (s) + \frac{1}{2} Cl_2 (g) \rightarrow LiCl (s)$ $-$ 409 kJ
$Cl (g) \rightarrow \frac{1}{2} Cl_2 (g)$ $-\frac{1}{2}$(243 kJ)
$Cl^- (g) \rightarrow Cl (g) + e^-$ $-$ (-349 kJ)

——————————————————————————

$Li^+ (g) + Cl^- (g) \rightarrow LiCl (s)$ $\Delta H = -$ 863 kJ

9.4   a. $Ba < Ca < Be$ b. $Ga < Ge < Se$ c. $Te < S < Cl$ d. $Bi < P < S$

9.5   C - Cl   C - H   C - Mg   C - O   C - S
   0.5   0.4   1.3   1.0   -
C - S < C - H < C - Cl < C - O < C - Mg

9.6

$$H - \overset{\displaystyle H}{\underset{\displaystyle H}{C}} - \overset{\displaystyle H}{\underset{\displaystyle H}{C}} - \overset{\cdot\cdot}{\underset{\cdot\cdot}{Cl}}:$$

9.7 $\overset{\cdot\cdot}{\underset{\cdot\cdot}{S}} = C = \overset{\cdot\cdot}{\underset{\cdot\cdot}{O}}$

9.8

$$\left[H - \overset{\displaystyle H}{\underset{\displaystyle H}{P}} - H \right]^{+}$$

9.9 $:\overset{\cdot\cdot}{O} = \overset{\cdot\cdot}{N} - \overset{\cdot\cdot}{\underset{\cdot\cdot}{Cl}}:$

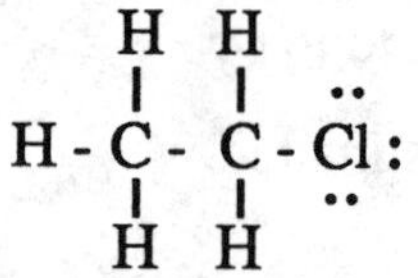

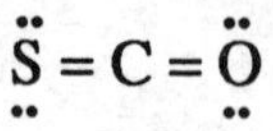

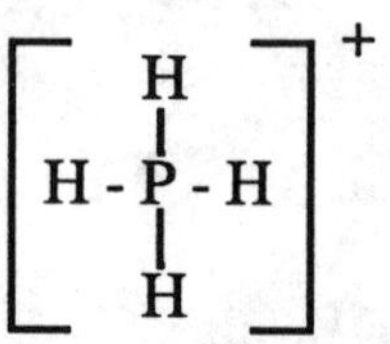

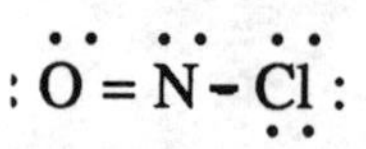

9.10 $$\left[\begin{array}{c} :\ddot{O}-N=\ddot{O} \\ | \\ :\ddot{O}: \end{array}\right]^{-} \longleftrightarrow \left[\begin{array}{c} :\ddot{O}-\ddot{O}: \\ \| \\ :\ddot{O}: \end{array}\right]^{-} \longleftrightarrow \left[\begin{array}{c} \ddot{O}=N-\ddot{O}: \\ | \\ :\ddot{O}: \end{array}\right]^{-}$$

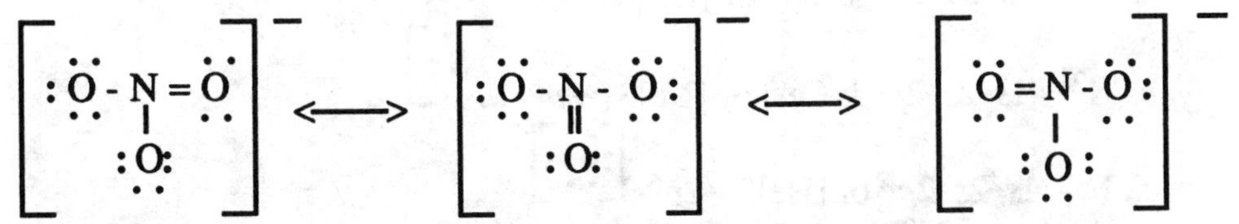

The resonance hybrid, which involves equal contributions from these three equivalent resonance structures, has N - to - O bonds with bond lengths and bond energies intermediate betweeen single and double bonds.

9.11 $:\ddot{F}-\ddot{C}l-\ddot{F}:$
$\qquad$ $|$
$\qquad$ $:\ddot{F}:$

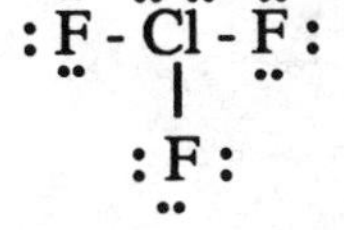

Estimation Exercises

9.1 a.
$$\begin{array}{ccccc} & H & & H & \\ & | & & | & \\ H- & C- & \dot{N}=\dot{N}- & C & -H \\ & | & & | & \\ & H & & H & \end{array}$$
N-to-N bond length = 123 pm

b. $:\ddot{F}:\ddot{O}:\ddot{F}:$ B.L. = $\frac{1}{2}$ (145 pm) + $\frac{1}{2}$ (143 pm) = 144 pm

9.2
ΔH bond breakage			ΔH bond formation	
1 mol CH	414 kJ		1 mol C - Cl	-339 kJ
1 mol Cl_2	243 kJ		1 mol H - Cl	-431 kJ
sum	657 kJ		sum	-770 kJ

$\Delta H = \Delta H_{breakage} + \Delta H_{formation}$
$\Delta H = 657$ kJ + (-770 kJ) = -113 kJ

Conceptual Exercises

9.1 a. Incorrect. The molecule ClO_2 has 19 valence electrons but the Lewis structure shows 20--one too many. b. Correct; c. Incorrect. The structure shown has an incomplete octet on a N atom. The correct structure is $:\ddot{F}-N=N-\ddot{F}:$

Review Questions

1. Group 8A, the noble gases.

4. a. Na $\cdot$ b. $:\ddot{O}:$ c. $:\ddot{F}:$ d. $\cdot Al \cdot$

5. a. $\cdot \dot{S}i \cdot$ b. Rb $\cdot$ c. $\cdot Ca \cdot$ d. $:\ddot{B}r \cdot$ e. $\cdot \ddot{A}s \cdot$

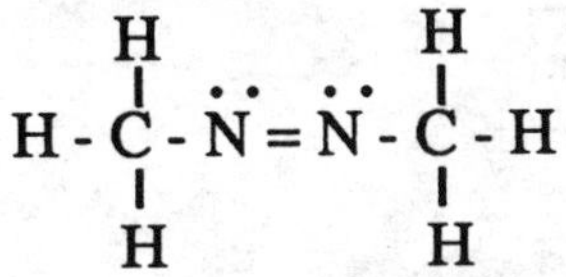

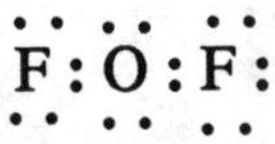

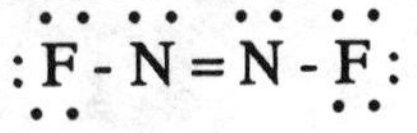

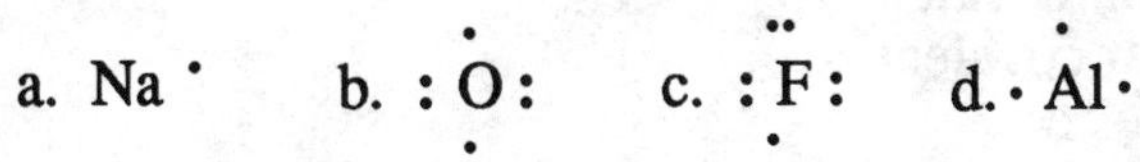

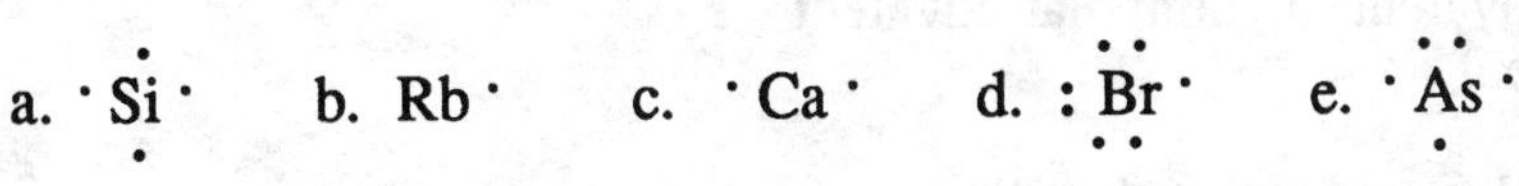

7. a. K^+ - $1s^22s^22p^63s^23p^6$ or [Ar], K^+

b. S^{2-} - $1s^22s^22p^63s^23p^6$ or $[Ne]3s^23p^6$, $[:\ddot{\underset{..}{S}}:]^{2-}$

c. F^- - $1s^22s^22p^6$ or $[He]2s^22p^6$, $[:\ddot{\underset{..}{F}}:]^-$

d. Al^{3+} - $1s^22s^22p^6$ or [Ne], Al^{3+}

8. a. Mg^{2+} - $1s^22s^22p^6$ or [Ne], Mg^{2+}

b. Cl^- - $1s^22s^22p^63s^23p^6$ or $[Ne]3s^23p^6$, $[:\ddot{\underset{..}{Cl}}:]^-$

c. Li^+ - $1s^2$ or [He], Li^+

d. N^{3-} - $1s^22s^22p^6$ or $[He]2s^22p^6$, $[:\ddot{\underset{..}{N}}:]^{3-}$

9. a. $\cdot Ca \cdot + 2\ \cdot\ddot{\underset{..}{Br}}: \longrightarrow [:\ddot{\underset{..}{Br}}:]^-\ Ca^{2+}\ [:\ddot{\underset{..}{Br}}:]^-$

b. $\cdot Mg \cdot + \cdot\ddot{\underset{..}{S}}\cdot \longrightarrow Mg^{2+}[:\ddot{\underset{..}{S}}:]^{2-}$

10. a. $2 \cdot\dot{Al}\cdot + :\dot{\underset{\cdot}{S}}: \longrightarrow 2\,[Al]^{3+} + 3[:\ddot{\underset{..}{S}}:]^{2-}$

b. $3\ \cdot Mg \cdot + 2\ \cdot\ddot{\underset{\cdot}{P}}\cdot \longrightarrow 3\,[Mg]^{2+} + 2[:\ddot{\underset{..}{P}}:]^{3-}$

12. 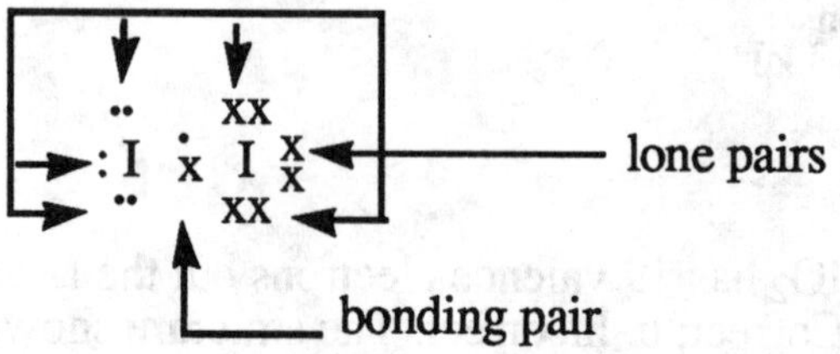

or $:\ddot{\underset{..}{I}}-\ddot{\underset{..}{I}}:$ where the dash line represents the bonding pair

13. $\delta+\ H\overset{x}{\cdot}\ddot{\underset{..}{F}}:\delta-$ or $\delta+\ H-\ddot{\underset{..}{F}}:\delta-$

14. a. ionic b. polar covalent
c. ionic d. polar covalent
e. ionic f. ionic
g. nonpolar covalent h. nonpolar covalent
i. polar covalent

16. a. N is more electronegative than S.
b. Cl is more electronegative than B.
c. F is more electronegative than As.
d. O is more electronegative than S.

17. a. F is more electronegative than Br.
     b. Br is more electronegative than Se.
     c. Cl is more electronegative than As.
     d. N is more electronegative than H.

21. a. 1   b. 4   c. 2   d. 1   e. 3   f. 1

23.

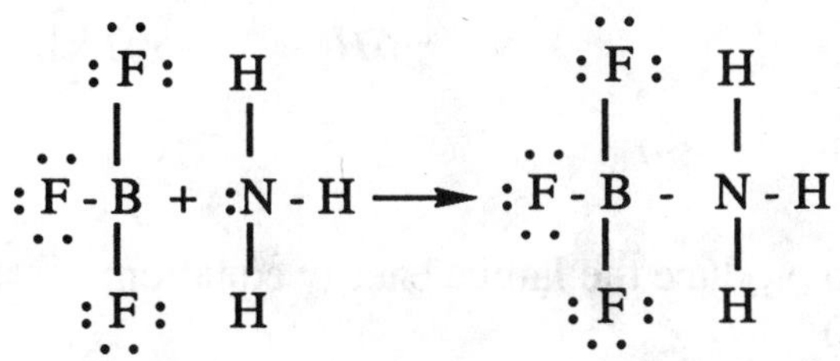

24. carbon, oxygen, nitrogen, and sulfur

25. carbon and nitrogen

31. (a), (b) and (c) are unsaturated.

32. (b) is an alkene
     (a) and (c) are alkynes

36. For addition polymerization, a monomer must have one or more multiple bonds in
     the molecule.

39.

Problems

41. a. Cr^{3+}- $1s^22s^22p^63s^23p^63d^3$ not noble gas
     b. Sc^{3+}- $1s^22s^22p^63s^23p^6$ noble gas
     c. Zn^{2+} - $1s^22s^22p^63s^23p^63d^{10}$ not noble gas
     d. Te^{2-}- $1s^22s^22p^63s^23p^63d^{10}4s^24p^64d^{10}5s^25p^6$ noble gas
     e. Zr^{4+}- $1s^22s^22p^63s^23p^63d^{10}4s^24p^6$ noble gas
     f. Cu^+ - $1s^22s^22p^63s^23p^63d^{10}$ not noble gas

43. Sn - [Kr] $4d^{10}5s^25p^2$
     Sn^{2+} - Kr] $4d^{10}5s^2$
     Sn^{4+}- [Kr] $4d^{10}$

45. a. K^+ + $[:\ddot{I}:]^-$ b. Ca^{2+} + $[:\ddot{O}:]^{2-}$

     c. Ba^{2+} + $2[:\ddot{Cl}:]^-$ d. $2 Rb^+$ + $[:\ddot{S}:]^{2-}$

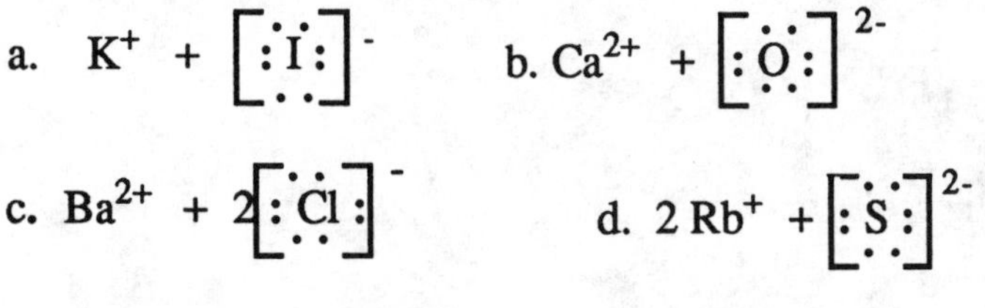

$$Na^+ (g) + F^- (g) \longrightarrow NaF (s)$$
$$Na (g) \longrightarrow Na^+ (g) + e^-$$
$$Na (s) \longrightarrow Na (g)$$
$$F (g) + e^- \longrightarrow F^- (g)$$
$$\frac{1}{2}F_2 (g) \longrightarrow F (g)$$

$\Delta H_{LE} = \quad -914 \text{ kJ}$
$\Delta H_{IE} = \quad 495 \text{ kJ}$
$\Delta H_{sub} = \quad 108 \text{ kJ}$
$\Delta H_{EA} = \quad -328 \text{ kJ}$
$\Delta H_{dis} = \frac{1}{2}(154 \text{ kJ})$

$$Na (s) + \frac{1}{2}F_2 (g) \longrightarrow NaF (s)$$

$\Delta H_f \quad = -562 \text{ kJ}$

Appendix D $\Delta H_f = -567.3 \text{ kJ}$

49. Equations must add together to produce the lattice energy equation:
$$Cs^+ (g) + Cl^- (g) \longrightarrow CsCl (s).$$

$$Cl^- (g) \longrightarrow Cl (g) + e^- \qquad\qquad - (-349 \text{ kJ/mol})$$
$$Cl (g) \longrightarrow \frac{1}{2} Cl_2 (g) \qquad\qquad - (\ 122 \text{ kJ/mol Cl})$$
$$\frac{1}{2} Cl_2 (g) + Cs (s) \longrightarrow CsCl (s) \qquad\qquad - \quad 442.8 \text{ kJ/mol}$$
$$Cs (g) \longrightarrow Cs (s) \qquad\qquad - (\ \ 77.6 \text{ kJ/mol})$$
$$Cs^+ (g) + e^- \longrightarrow Cs (g) \text{ (ionization energy table 8.2)} \qquad - (\ 375.7 \text{ kJ/mol})$$

$$Cs^+ (g) + Cl^- (g) \longrightarrow CsCl (s) \qquad\qquad \Delta H° = \text{lattice energy}$$
$\Delta H° = 349 \text{ kJ} - 122 \text{ kJ} - 442.8 \text{ kJ} - 77.6 \text{ kJ} - 375.7 \text{ kJ}$
$\Delta H° = -669 \text{ kJ}$

51. a.
```
        ..
    H - P - H
        |
        H
```
b.
```
          :F:
      ..   |   ..
    :F -  C - F:
      ..   |   ..
          :F:
          ..
```

53. a. $B < N < F$ \qquad b. $Ca < As < Br$ \qquad c. $Ga < C < O$

55.
a.
$$\overset{\delta+\ \ \delta-}{F - F} < \overset{\delta+\ \delta-}{Cl - F} < \overset{\delta+\ \delta-}{Br - F} < \overset{\delta+\ \delta-}{I - F} < H - F$$
$$\ \ 4.0 \quad\ 3.0\ \ 4.0 \ \ 2.8\ \ 4.0\ \ 2.5\ \ 4.0\ \ 2.1\ \ 4.0$$

b.
$$H - H < \overset{\delta+\ \ \delta-}{H - I} < \overset{\delta+\ \delta-}{H - Br} < \overset{\delta+\ \ \delta-}{H - Cl} < \overset{\delta+\ \delta-}{H - F}$$
$$\ \ 2.1 \qquad 2.1\ \ 2.5\ \ 2.1\ \ 2.8 \quad 2.1\ \ 3.0\ \ 2.1\ \ 4.0$$

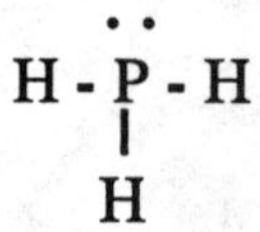

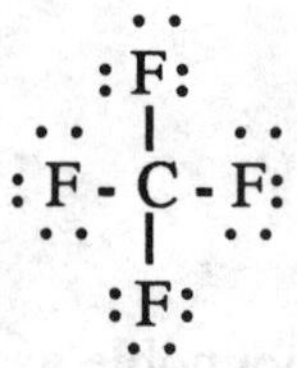

57. a. $B.L. (I - Cl) = \frac{1}{2} B.L. (I - I) + \frac{1}{2} B.L. (Cl - Cl)$

$B.L. = \frac{1}{2}(266\ pm) + \frac{1}{2}(199\ pm)$

$B.L. = 233\ pm$

b. $B.L. (C - F) = \frac{1}{2} B.L. (C - C) + \frac{1}{2} B.L. (F - F)$

$B.L. = \frac{1}{2}(154\ pm) + \frac{1}{2}(143\ pm)$

$B.L. = 149\ pm$

59. $\Delta H = BE (H_2) + BE (F_2) - 2BE (HF)$
$\Delta H = 436\ kJ + 159\ kJ - 2\ mol \times 565\ kJ/mol$
$\Delta H = -535\ kJ$

61. a. 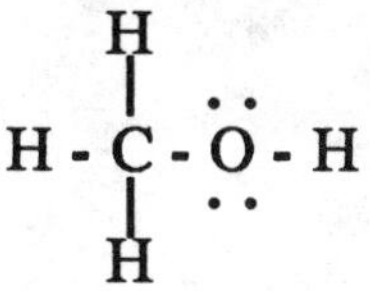 b.  c. 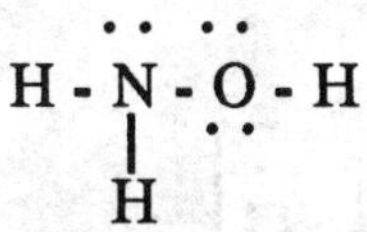

d.  e. 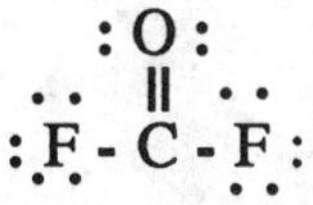 f. 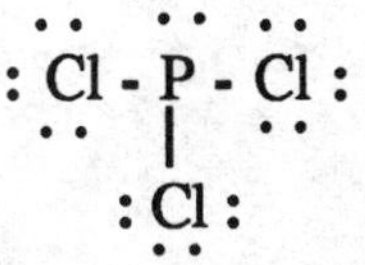

63. a. $[:C = N = O:]^-$

$C\ 4 - \frac{1}{2}(4) - 4 = -2$

$N\ 5 - \frac{1}{2}(8) = 1$

$O\ 6 - \frac{1}{2}(4) - 4 = 0$

b. $[:N \equiv C - O:]^-$

$N\ 5 - \frac{1}{2}(6) - 2 = 0$

$C\ 4 - \frac{1}{2}(8) = 0$

$O\ 6 - \frac{1}{2}(2) - 6 = -1$

(b) is preferred since the total formal charge is less.

65. 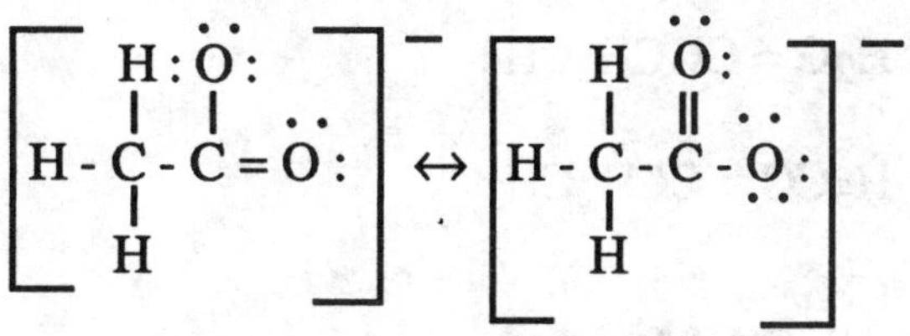

The resonance hybrid is a combination of these two structures, where the C-to-O bonds are intermediate between single and double bonds, and the carbon and two oxygen atoms share a pair of delocalized electrons.

67.
$$\left[:\overset{..}{O}-\overset{..}{C}=\overset{..}{O}:\right]^{2-} \longleftrightarrow \left[\overset{..}{O}=\overset{..}{C}-\overset{..}{O}:\right]^{2-} \longleftrightarrow \left[:\overset{..}{O}-\overset{..}{C}-\overset{..}{O}:\right]^{2-}$$

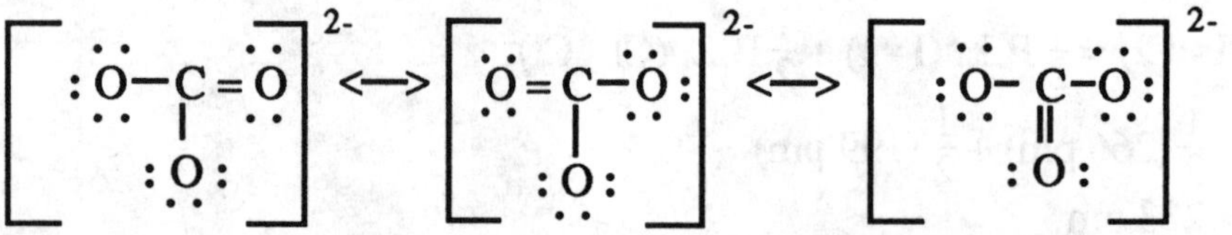

The resonance hybrid combines all three structures, so that the carbon and all three oxygen atoms share a pair of delocalized electrons.

69.   a. $\cdot \overset{..}{N} = \overset{..}{O}$ b. $:\overset{..}{Cl}-\overset{}{B}-\overset{..}{Cl}:$ c. $:\overset{..}{O}-\overset{..}{Cl}-\overset{..}{O}\cdot$ d. $H-\overset{H}{\underset{H}{C}}-Be-\overset{H}{\underset{H}{C}}-H$

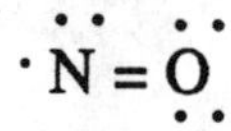

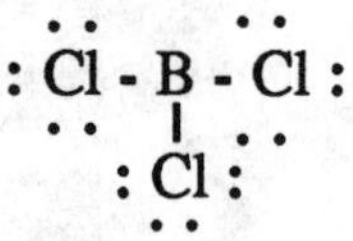

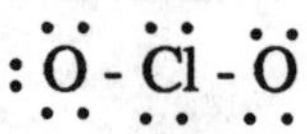

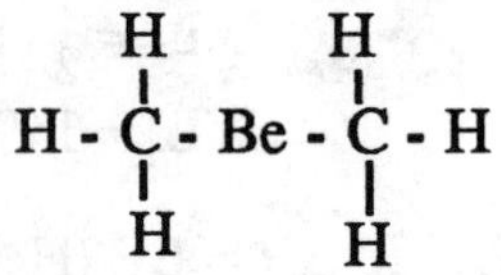

NO and ClO_2 are odd-electron molecules; BCl_3 and $Be(CH_3)_2$ have incomplete octets.

71.   a. $\left[:\overset{..}{I}\cdot\overset{..}{I}\cdot\overset{..}{I}:\right]^-$

b. $\left[:\overset{..}{F}-\overset{\overset{\overset{..}{F}:}{|}}{\underset{\underset{:\overset{..}{F}:}{|}}{I}}-\overset{..}{F}:\right]^-$

c. $\left[:\overset{..}{Cl}-\overset{..}{I}-\overset{..}{Cl}:\right]^-$

d. $\left[:\overset{..}{F}-\overset{\overset{:\overset{..}{F}: \quad \overset{..}{F}:}{}}{\underset{\underset{:\overset{..}{F}:}{|}}{S}}-\overset{..}{F}:\right]^-$

73.   a. $\left[:\overset{..}{O}:H\right]^-$ b. $[:C\equiv N:]^-$

c. $\left[:\overset{..}{O}=N-\overset{..}{O}:\right]^- \longleftrightarrow \left[\overset{..}{O}-N=\overset{..}{O}:\right]^-$

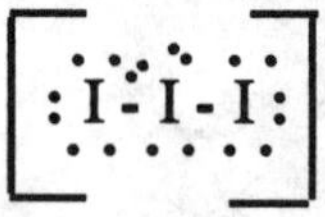

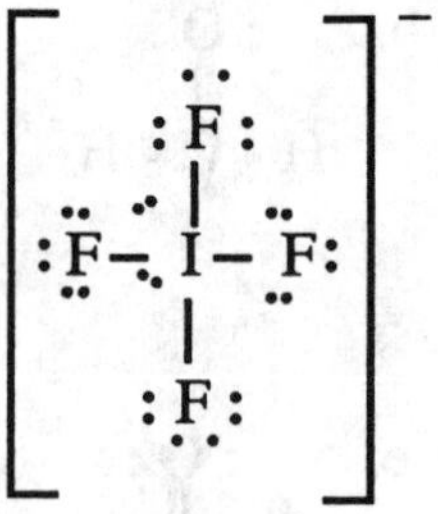

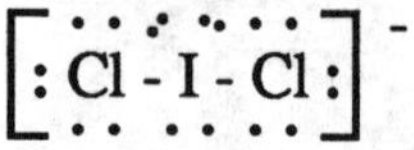

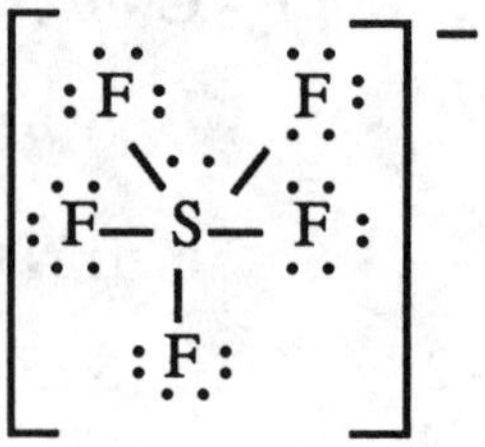

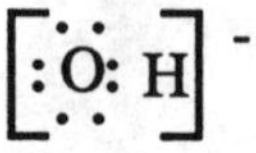

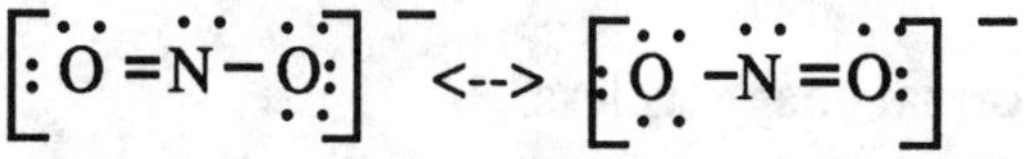

75.   a. $H_2C = CH_2$ b. $H_2C = CHCH_2CH_3$

c. $HC \equiv CCH_3$ d. $H_3CC \equiv CCH_2CH_3$

77.   a. $H_2C = CH_2 + H_2 \longrightarrow H_3C - CH_3$
b. $H_2C = CHCH_3 + Br_2 \longrightarrow H_2C - CHCH_3$
                                                            $\underset{Br}{|} \quad \underset{Br}{|}$
c. $HC \equiv CH + 2H_2 \longrightarrow H_3C - CH_3$

79. a. $-[-CH_2CH_2CH_2CH_2CH_2CH_2CH_2CH_2-]-$

b.
$$-[-CH-CH-CH-CH-CH-CH-CH-CH-]-$$
$$\quad\ |\quad\ |\quad\ |\quad\ |\quad\ |\quad\ |\quad\ |\quad\ |$$
$$\quad Cl\ \ H\ \ Cl\ \ H\ \ Cl\ \ H\ \ Cl\ \ H$$

c.
$$-[-CH-CH-CH-CH-CH-CH-CH-CH-]-$$
$$\quad\ \ |\quad\ |\quad\quad\ |\quad\ |\quad\quad\ |\quad\ |\quad\quad\ |\quad\ |$$
$$\quad CH_3\ H\quad CH_3\ H\quad CH_3\ H\quad CH_3\ H$$

81.
W	X	Y	Z
7A	1A	2A	6A

:Ẅ: X·   ·Y·   :Ż:

1-   1+   2+   2-

83.
$$\begin{array}{ccc}
H & & H \\
| & \cdot\cdot & | \\
H-C-O-C-H \\
| & \cdot\cdot & | \\
H & & H
\end{array}$$
dimethyl ether

$$\begin{array}{ccc}
H & H & \\
| & | & \\
H-C-C-\ddot{O}-H \\
| & | & \\
H & H &
\end{array}$$
ethanol

85.
$$\begin{array}{cccc}
H & H & H & H \\
| & | & | & | \\
C=C-C=C \\
| & & & | \\
H & & & H
\end{array}$$

$$\begin{array}{cccc}
H & & H & H \\
| & & | & | \\
C=C=C-C-H \\
| & & & | \\
H & & & H
\end{array}$$

$$\begin{array}{cccc}
& & H & H \\
& & | & | \\
H-C\equiv C-C-C-H \\
& & | & | \\
& & H & H
\end{array}$$

$$\begin{array}{cccc}
H & & & H \\
| & & & | \\
H-C-C\equiv C-C-H \\
| & & & | \\
H & & & H
\end{array}$$

$$\begin{array}{cc}
H & H \\
| & | \\
C-C-H \\
|| & | \\
C-C-H \\
| & | \\
H & H
\end{array}$$

87. :F̈-S̈-S̈-F̈:   :F̈-S̈=S̈:

$$\qquad\qquad\qquad\qquad\qquad |$$
$$\qquad\qquad\qquad\qquad\quad :F̈:$$

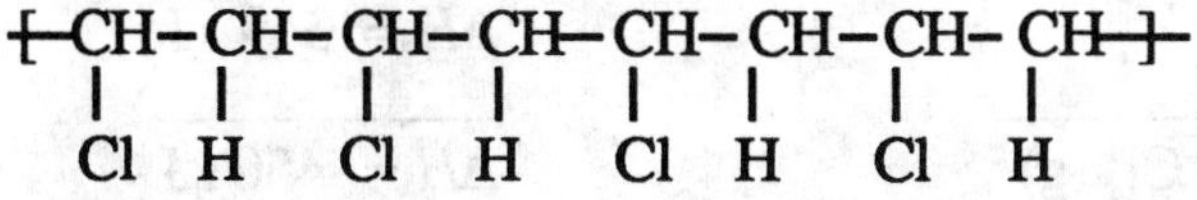

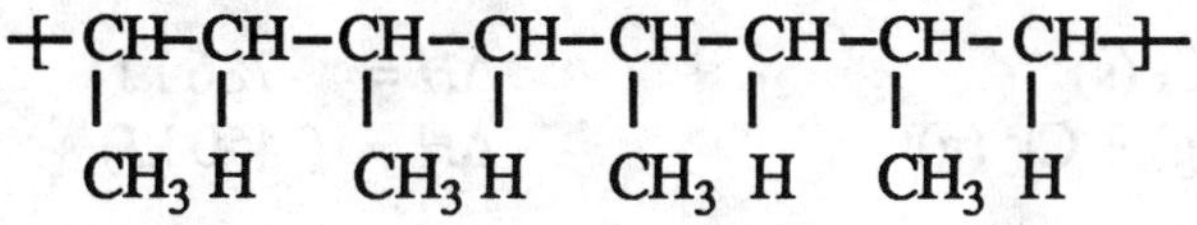

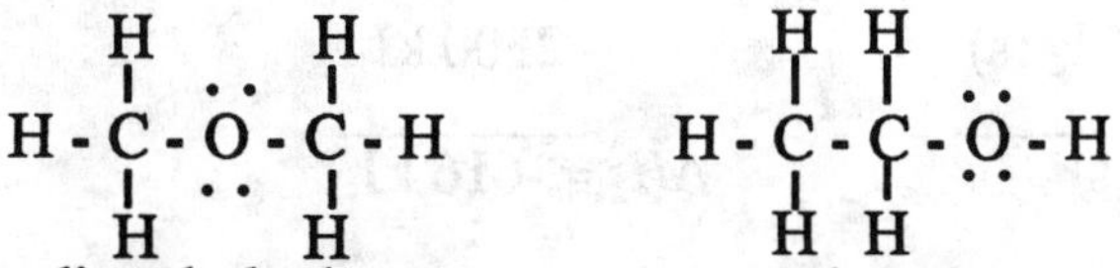

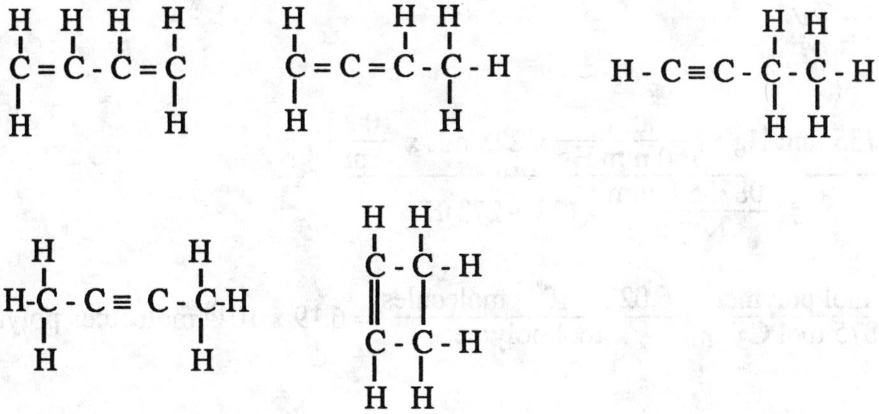

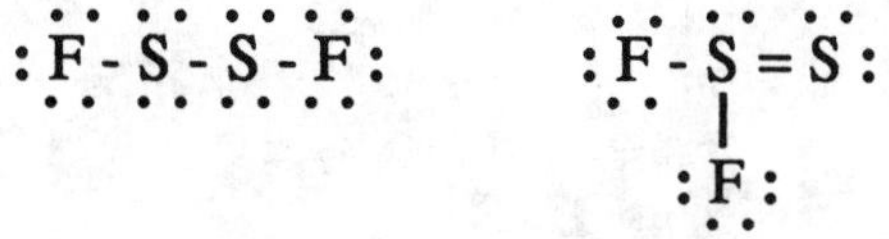

89.

$Na(g) + Cl(g) \rightarrow Na^+Cl^-(g)$	$\Delta H = -304$ kJ
$Na^+(g) + e^- \rightarrow Na(g)$	$\Delta H = -495$ kJ
$Cl^-(g) \rightarrow Cl(g) + e^-$	$\Delta H = 349$ kJ
$Na^+(g) + Cl^-(g) \rightarrow Na^+Cl^-(g)$	$\Delta H = -450$ kJ

$Na^+(g) + Cl^-(g) \rightarrow NaCl(s)$	$\Delta H = -786$ kJ
$Na^+(g)\,Cl^-(g) \rightarrow Na^+(g) + Cl^-(g)$	$\Delta H = -(-450$ kJ$)$
$Na^+Cl^-(g) \rightarrow NaCl(s)$	$\Delta H = -336$ kJ

91.

	ΔH_f
$Mg(s) + Cl_2(g) \longrightarrow MgCl_2(s)$	
$Mg(s) \longrightarrow Mg(g)$	150 kJ
$Mg(g) \longrightarrow Mg^+(g) + e^-$	738 kJ
$Mg^+(g) \longrightarrow Mg^{2+}(g) + e^-$	1451 kJ
$Cl_2(g) \longrightarrow 2\,Cl(g)$	243 kJ
$2\,Cl(g) + e^- \longrightarrow 2\,Cl^-$	2 mol x -349 kJ/mol
$Mg^{2+}(g) + 2\,Cl^-(g) \longrightarrow MgCl_2(s)$	- 2500 kJ
$Mg(s) + Cl_2(g) \longrightarrow MgCl_2(s)$	$\Delta H_f = -616$ kJ

$MgCl_2$ is much more stable than MgCl because of the additional energy given off during its formation.

93. $n = \dfrac{PV}{RT}$

$$\frac{735 \text{ mm Hg} \times \dfrac{\text{atm}}{760 \text{ mm Hg}} \times 225 \text{ mL} \times \dfrac{10^{-3}\text{ L}}{\text{mL}}}{\dfrac{.08206 \text{ L atm}}{\text{K mol}} \times (22 + 273)\text{K}} \times$$

$$\frac{\text{mol polymer}}{875 \text{ mol } C_3H_6} \times \frac{6.022 \times 10^{23} \text{ molecules}}{\text{mol polymer}} = 6.19 \times 10^{18} \text{ molecules polymer}$$

Chapter 10

Bonding Theory and Molecular Structure

Exercises

10.1 $SbCl_5$

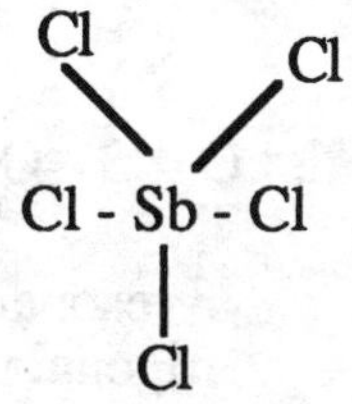

The five bonding pairs of electrons (AX_5) produce both a trigonal bipyramidal electron-group geometry and molecular geometry.

10.2 There are 5 pairs of electrons, but one pair is a lone pair of electrons (AX_4E). That pair will be one of the pairs on the equitorial plane of the structure. This produces a seesaw shape molecule.

10.3

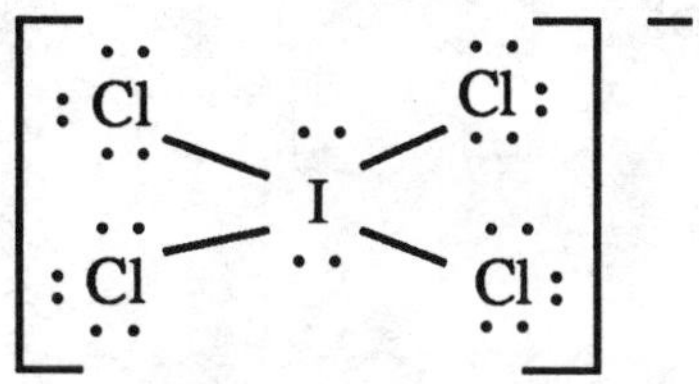

There are 6 electron pairs around the I atom but two are lone pairs (AX_4E_2). The Cl atoms are coplanar about the central I atom, and the lone pairs of electrons are situated above and below the plane so the shape is square planar.

10.4

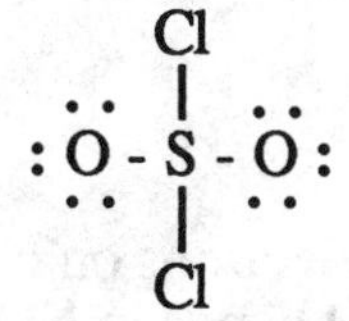

There are four bonding pairs and no lone pairs of electrons around the S atom, (AX_4), so the shape is tetrahedral.

10.5

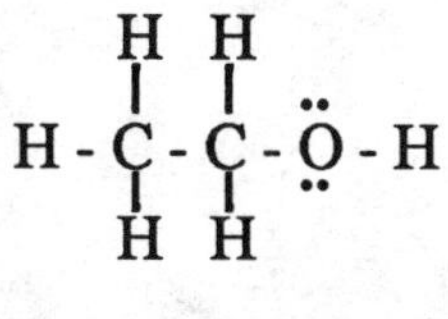

For both carbon atoms there are 4 bonding pairs and no lone pairs of electrons, (AX_4), so the shape around each carbon atom is tetrahedral. Around the O atom there are four electron pairs but two of them are lone pairs, (AX_2E_2), so the COH bond is “bent.”

10.6 a.

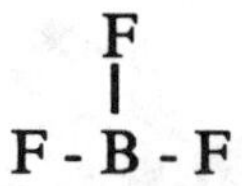

trigonal planar
symmetric
nonpolar

b.

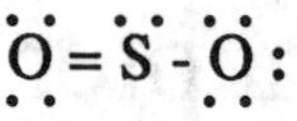

bent
nonsymmetric
polar

c.

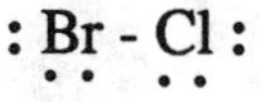

linear
different electronegativities
polar

d.

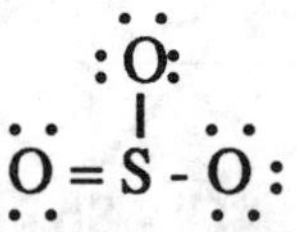

3 resonance structures
trigonal planar
symmetric
nonpolar

10.7

VSEPR notation: AX_2E_3

electron-group geometry: trigonal bipyramidal

linear shape

hybridization scheme: sp^3d

I [Kr] $4d^{10}$ 

sp^3d

Conceptual Exercises

10.1 NOF is probably 110° and NO_2F is 118°. The lone pair of electrons on N in NOF is more repulsive than are the bonding pairs in the N - to - O bonds in NO_2F.

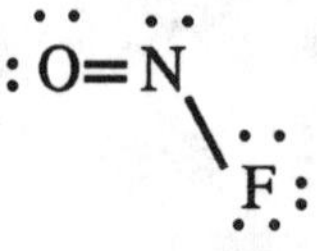

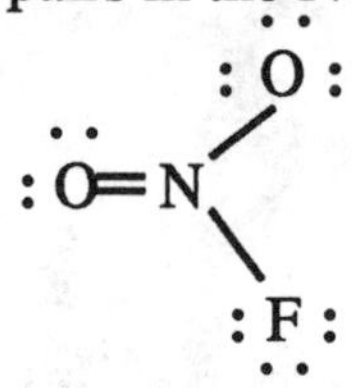

10.2
$$
\begin{array}{c}
\quad\quad \text{H} \\
\quad\quad | \\
\text{H} - \text{C} - \ddot{\text{O}} - \text{H} \\
\quad\quad | \\
\quad\quad \text{H}
\end{array}
$$

     a. Around the C is tetrahedral.

       Around the O is bent.

     b. C is sp^3, O is sp^3

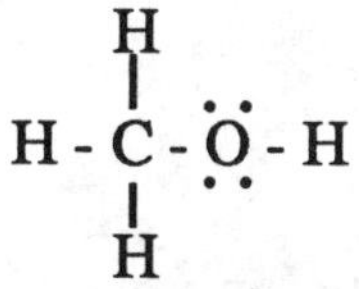

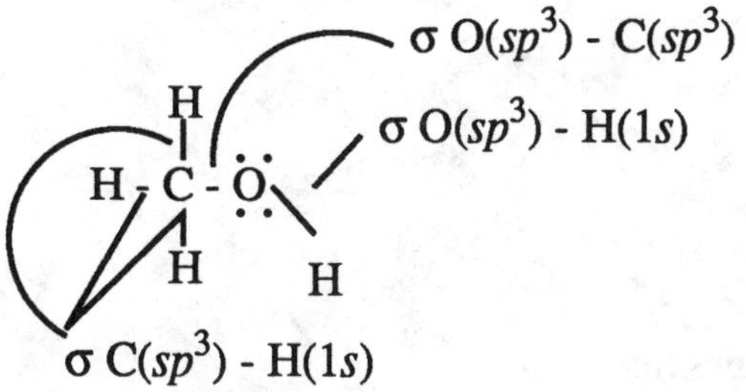

Review Questions

6.   a. AX_3E
$$
\begin{array}{c}
\text{H} - \ddot{\text{P}} - \text{H} \\
\quad | \\
\quad \text{H}
\end{array}
$$

    b. AXE_3 $\left[:\ddot{\text{O}} - \ddot{\text{C}}\text{l} : \right]^-$

    c. AX_2E $\left[\ddot{\text{O}} = \dot{\text{N}} - \ddot{\text{O}}: \right]^- \longleftrightarrow \left[:\ddot{\text{O}} - \dot{\text{N}} = \ddot{\text{O}} \right]^-$

7.   a. 180° $:\ddot{\text{C}}\text{l} - \text{Be} - \ddot{\text{C}}\text{l} :$

    b. 120° $\ddot{\text{O}} = \ddot{\text{S}} - \ddot{\text{O}} :$

    c. 109.5° $\text{H} - \ddot{\text{O}} - \text{H}$

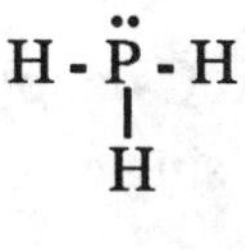

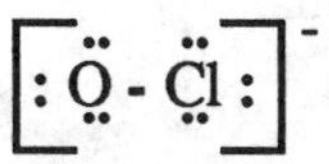

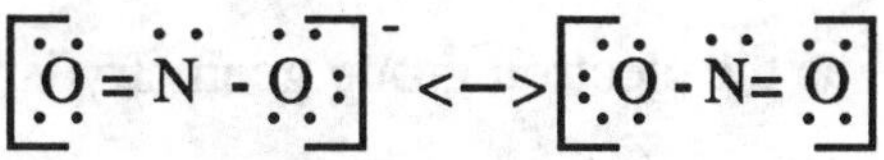

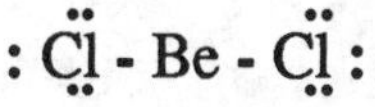

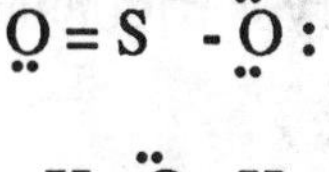

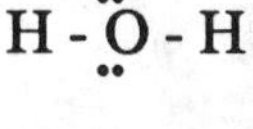

8.   Yes, if the molecule is diatomic. For example, HCl, which has the VSEPR notation AXE_3 on Cl.

11.   Nonpolar covalent bonds, bonds between two atoms of the same element, do not have a dipole moment. Similarly, a bond between two atoms of the same electronegativity, for example C and S, is nonpolar.

12.   Molecules, even those with polar bonds, may be nonpolar if the dipole moments cancel out; that is, if they are of equal magnitude and act in opposition to each other, as in CCl_4.

13.　$\ddot{O} = \ddot{S} - \ddot{O}$　　　$\ddot{O} = S - \ddot{O}$

$\underset{\displaystyle :\ddot{O}:}{\big|}$

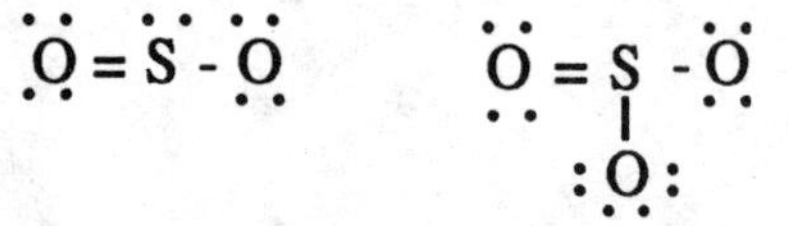

SO$_2$ has a bent shape with polar bonds; it is a polar molecule. The dipole moments in SO$_3$, however, cancel out because SO$_3$ has a trigonal planar shape.

18.　trigonal planar sp^2
　　　octahedral sp^3d^2 (or d^2sp^3)

19.　a.　$: Cl - \overset{\displaystyle :\ddot{Cl}:}{\underset{\displaystyle :\ddot{Cl}:}{C}} - Cl :$　　4 bonding pairs
tetrahedral
sp^3

　　　b.　$\ddot{O} = C = \ddot{O}$　　　2 bonding pairs linear sp

23.　All of the bonds in an alkane molecule are single bonds and are thus σ bonds.

24.　No. An alkene hydrocarbon has one or more C-to-C double bonds, each consisting of a σ bond and a π bond. All C-to-C single bonds in the alkene hydrocarbon are σ bonds as are all C-to-H bonds.

Problems

29.　a.　$H - \ddot{O} - H$
　　　　Linear is improbable, as the electron group geometry is tetrahedral.

　　　b.　$\ddot{O} = S - \ddot{O}$

$:O:$

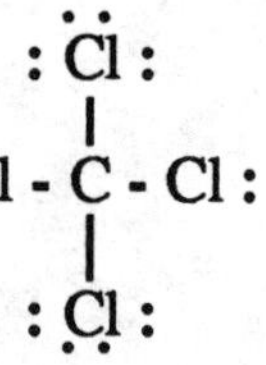

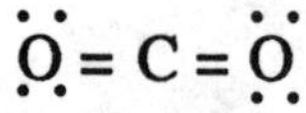

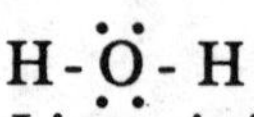

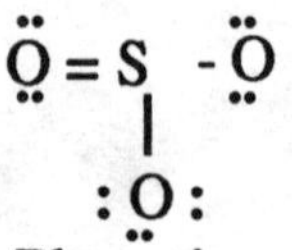

　　　　Planar is probable, as electron group geometry is trigonal planar.

　　　c.　$H - \ddot{P} - H$

$\underset{\displaystyle H}{\big|}$

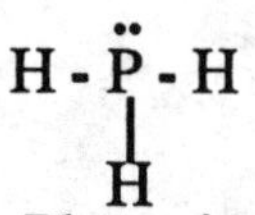

　　　　Planar is not probable, as electron group geometry is tetrahedral.

31.　$: \ddot{F} - \overset{}{\underset{\displaystyle :\ddot{F}:}{B}} - \ddot{F} :$　　　$: \ddot{F} - \overset{}{\underset{\displaystyle :\ddot{F}:}{\ddot{Cl}}} - \ddot{F} :$

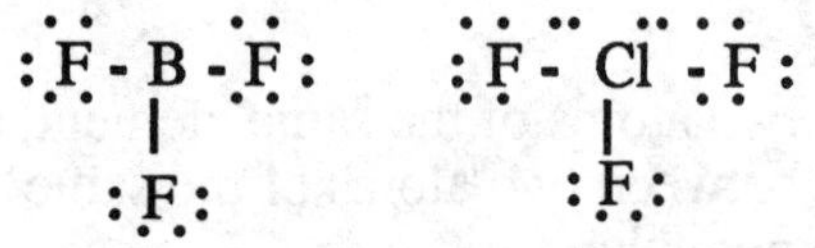

BF$_3$ has 3 electron groups, all bonding pairs, and so BF$_3$ is trigonal planar. ClF$_3$ has 3 bonding pairs and 2 lone pairs. The electron-group geometry is trigonal bipyramidal, and the molecular geometry is T-shaped.

33. a. $:\overset{..}{C}l-P-\overset{..}{C}l:$ 4 electron-groups   b. $\left[\,:\overset{..}{O}-\overset{\overset{\displaystyle :\overset{..}{O}:}{|}}{\underset{\underset{\displaystyle :\overset{..}{O}:}{|}}{Cl}}-\overset{..}{O}:\,\right]^{-}$ 4 electron-groups
                 $\overset{|}{:\overset{..}{C}l:}$ 3 bond pair   4 bond pair
                         1 lone pair   tetrahedral
                         pyramidal

c. $:\overset{\overset{\displaystyle :\overset{..}{F}:}{|}}{\underset{\underset{\displaystyle :\overset{..}{F}:}{|}}{\overset{..}{F}-Xe}}-\overset{..}{F}:$ 6 electron-groups   d. $\left[\,:\overset{..}{O}-C\equiv N:\,\right]^{-}$ 2 electron-groups
                         4 bond pair   both bonding
                         2 lone pair   linear
                         square planar

35. a. $H-\overset{..}{\underset{..}{O}}-\overset{..}{\underset{..}{O}}-H$ b. $H-\overset{..}{\underset{..}{O}}-C\equiv N:$

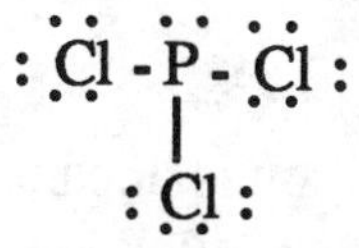

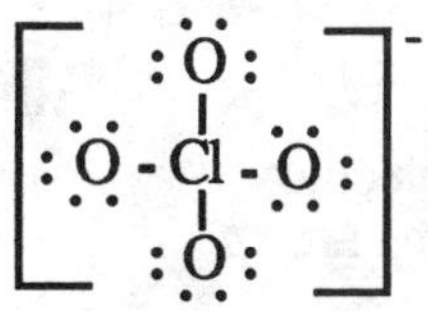

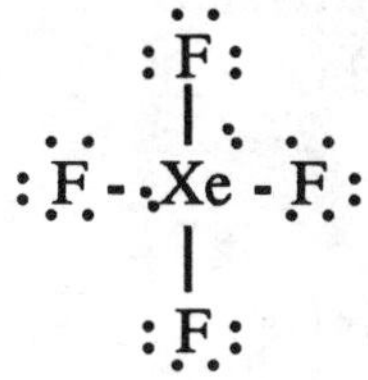

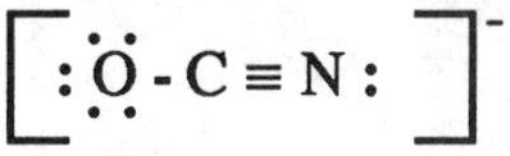

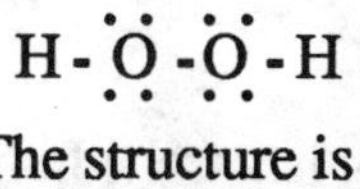

The structure is "bent" at   The shape around the
the O-O-H bonds. The   C is linear but bent
overall shape is a nonplanar   around the O.
"zigzag."

c. $H-\overset{..}{\underset{..}{O}}-\overset{..}{N}=\overset{..}{O}:$ The O-N-O portion of the molecule is a bent shape from an
                                         electron-group geometry of trigonal planar, but the O-H
                                         bond is bent out of the plane.

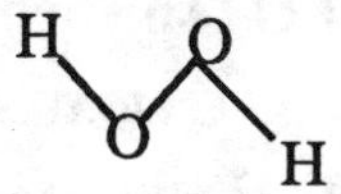

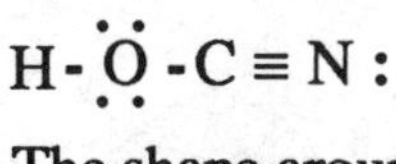

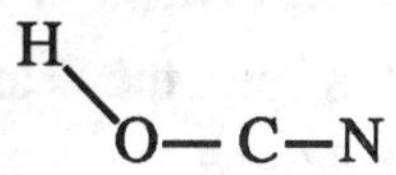

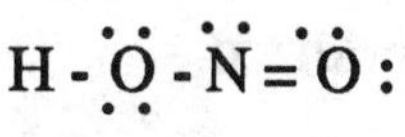

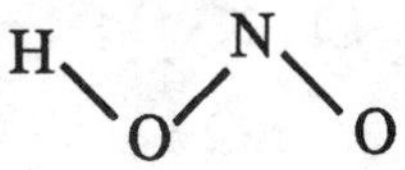

37. a. $:\overset{..}{\underset{..}{F}}-\overset{..}{\underset{..}{O}}-\overset{..}{\underset{..}{F}}:$ b. $\left[\begin{array}{c} H \\ | \\ H-N-H \\ | \\ H \end{array}\right]^{+}$
          tetrahedral sp^3

                                                       tetrahedral sp^3

c. $\overset{..}{\underset{..}{O}}=C=\overset{..}{\underset{..}{O}}$ d. $:\overset{..}{\underset{..}{C}l}-\overset{\overset{\displaystyle :O:}{||}}{C}-\overset{..}{\underset{..}{C}l}:$

     linear sp trigonal planar sp^2

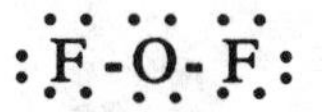

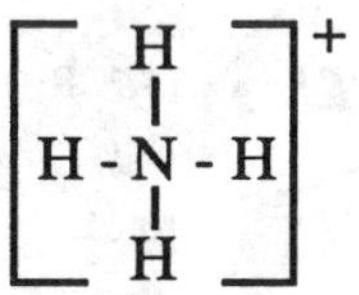

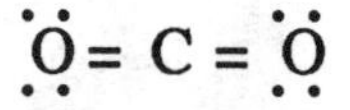

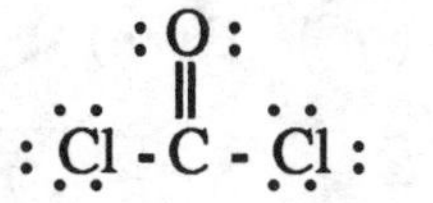

39.   a. : N ≡ C - C ≡ N :

A linear molecule; both C atoms are *sp* hybridized.

b. H - N̈ = C = Ö

The N-C-O portion is linear; the C atom is *sp* hybridized. The H-N-C portion is trigonal planar; the N is sp^2 hybridized.

c.
$$\begin{array}{c} H \\ | \\ H - \underset{\cdot\cdot}{N} - \underset{\cdot\cdot}{O} - H \end{array}$$

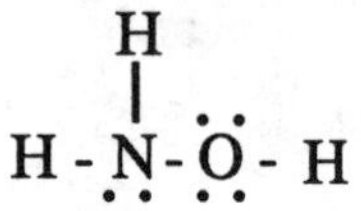

The NH$_2$-O portion is pyramidal; the N atom is sp^3 hybridized. The N-O-H portion is bent; the O atom is sp^3 hybridized.

d.
$$\begin{array}{c} H\;\;:O: \\ |\;\;\;\; || \\ H - C - C - \underset{\cdot\cdot}{O} - H \\ | \\ H \end{array}$$

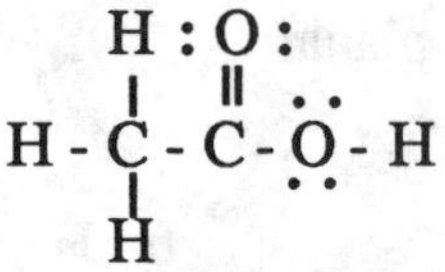

The CH$_3$-C portion is tetrahedral; the first C atom is sp^3 hybridized. The C-CO-O portion is trigonal planar; the second C atom is sp^2 hybridized. The C-O-H portion is bent; the central O atom is sp^3 hybridized.

41.   a. : C̈l - N - Ö : <—> : C̈l - N = Ö   trigonal planar

σ: Cl (3*p*) - N (sp^2) σ: N (sp^2) - O (2*p*)

Cl — N — O

σ: N (sp^2) - O (2*p*) O π: N (2*p*) - O (2*p*)

b. : F̈ - Ö - F̈ : σ: O (sp^3) - F (2*p*)   O   σ: O (sp^3) - F (2*p*)

F   F

c.
$$\left[:\overset{\cdot\cdot}{O} - C = \overset{\cdot\cdot}{O} \atop :O: \right]^{2-} <-> \left[:\overset{\cdot\cdot}{O} - C - \overset{\cdot\cdot}{O}: \atop :O: \right]^{2-} <-> \left[\overset{\cdot\cdot}{O} = C - \overset{\cdot\cdot}{O}: \atop :O: \right]^{2-}$$

trigonal planar   O   O

σ: C (sp^2) - O (2*p*) C σ: C (sp^2) - O (2*p*)

σ: C (sp^2) - O (2*p*) O π: C (2*p*) - O (2*p*)

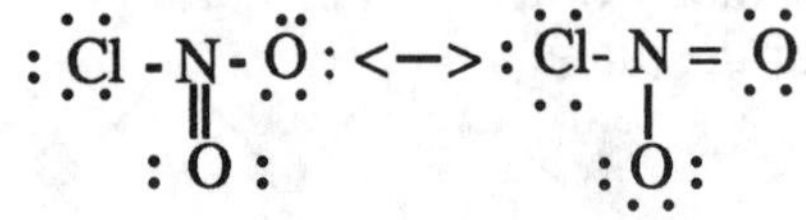

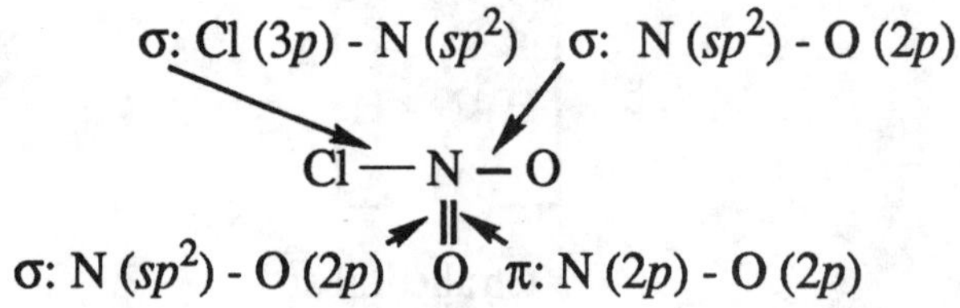

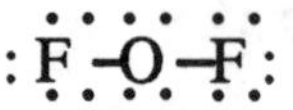

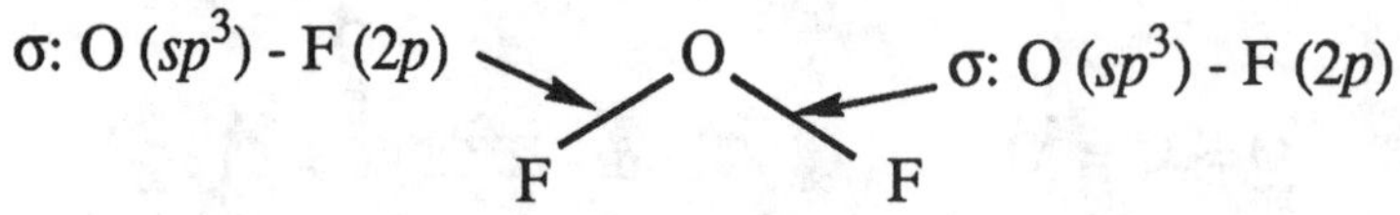

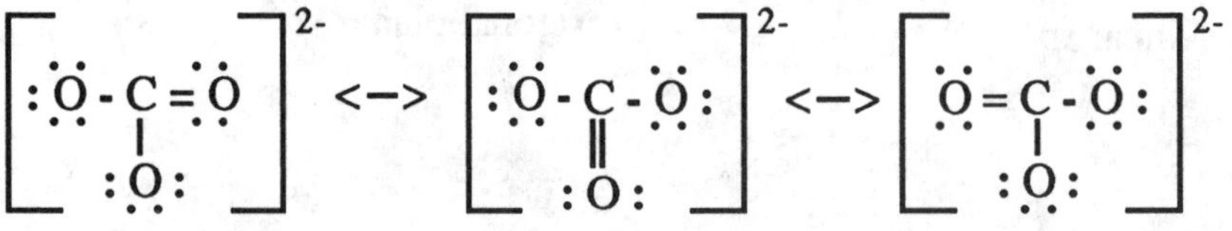

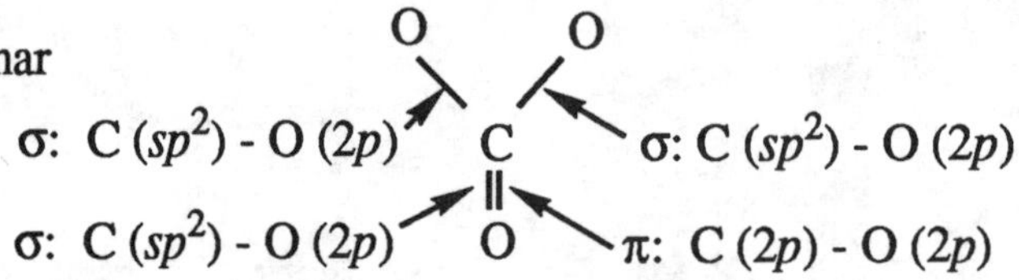

43. $\left[:\ddot{C}l - \ddot{I} - \ddot{C}l: \right]^{-}$ $\left[:\ddot{C}l - \ddot{I} - \ddot{C}l: \right]^{+}$

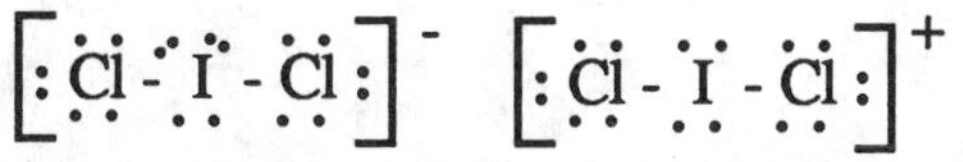

The electron-group geometry in the ICl_2^{-} ion is that of AX_2E_3 trigonal bipyramidal (the ion is linear). The hybridization scheme is sp^3d. The electron-group geometry in the ICl_2^{+} ion is that of AX_2E_2 (the ion is bent). The hybridization scheme is sp^3.

45.

$$\left[\begin{array}{cc} :O: & :O: \\ \| & \\ C & C \\ | & \\ :O: & :O: \end{array} \right]^{2-} <-> \left[\begin{array}{cc} :O: & :\ddot{O}: \\ \| & \\ C & C \\ | & \| \\ :\ddot{O}: & :O: \end{array} \right]^{2-}$$

$$\wedge \mathrel{\rlap{|}\vee} \qquad\qquad \wedge \mathrel{\rlap{|}\vee}$$

$$\left[\begin{array}{cc} :\ddot{O}: & :O: \\ | & \| \\ C & C \\ \| & | \\ :O: & :\ddot{O}: \end{array} \right]^{2-} <-> \left[\begin{array}{cc} :\ddot{O}: & :\ddot{O}: \\ | & | \\ C & C \\ \| & \| \\ :O: & :O: \end{array} \right]^{2-}$$

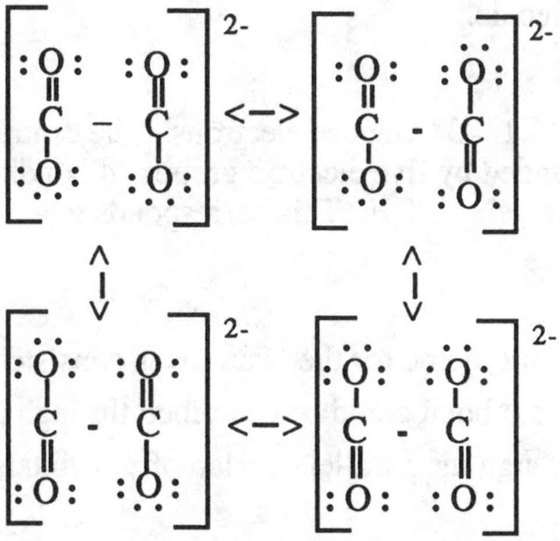

The 157 pm is the length for a single bond between the carbon atoms. The 125 pm is not short enough for a double bond between C and O, but it would be correct for a delocalized pair of electrons shared by two O and C. Repulsions between lone pair electrons on the O atoms bonded to the same C atom cause the O-C-O bonds to open up to a larger angle (126°) than the 120° angles corresponding to sp^2 hybridization. The C atoms are sp^2 hybridized.

47. (a) and (c) can be cis trans isomers.

a.

<!-- cis and trans structures -->

c.

<!-- cis and trans structures -->

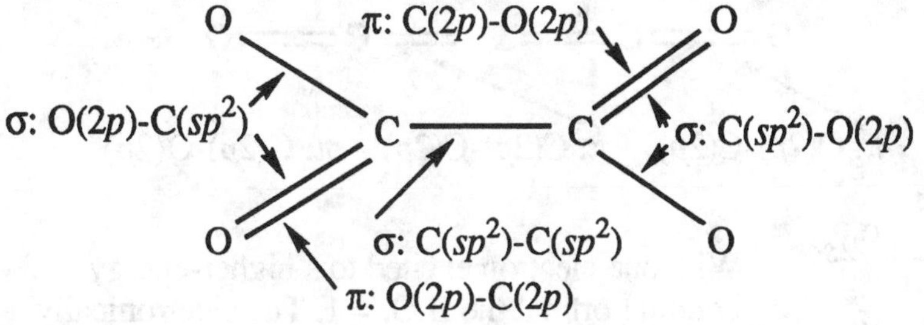

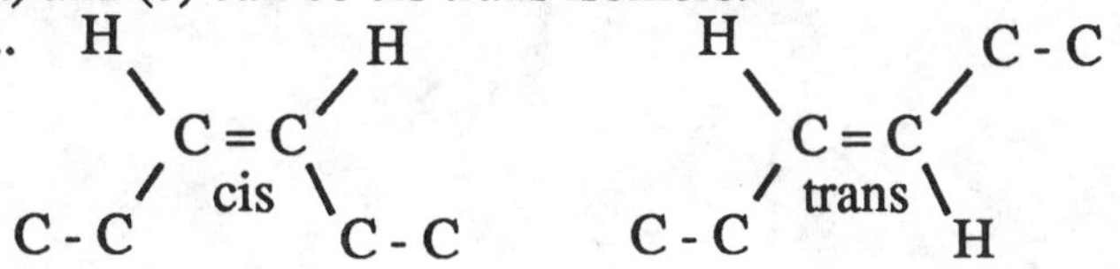

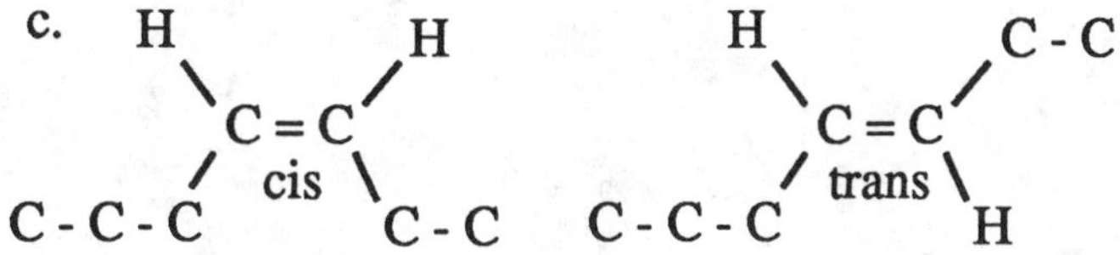

49. $\underline{\uparrow\downarrow}$ σ_{1s}^*
     With the antibonding orbital filled there is no net attraction
     to hold the atoms together. That is, the bond order, which

     $\underline{\uparrow\downarrow}$ σ_{1s}
     is one-half the difference between the number of electrons in
     bonding and in antibonding orbitals, is zero. There is no bond.

51. No. Although the statement is correct for diatomic molecules, for molecules with
     more than two atoms the dipole moments of different bonds may cancel each
     other, resulting in a nonpolar molecule.

53.
$$\left[\begin{array}{c} :\ddot{F}: \\ | \\ :\ddot{F}-\ddot{Br}-\ddot{F}: \\ | \\ :\ddot{F}: \end{array} \right]^{+}$$

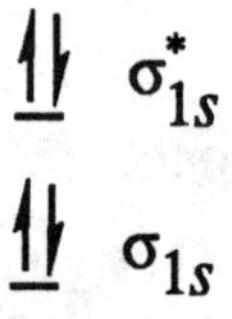

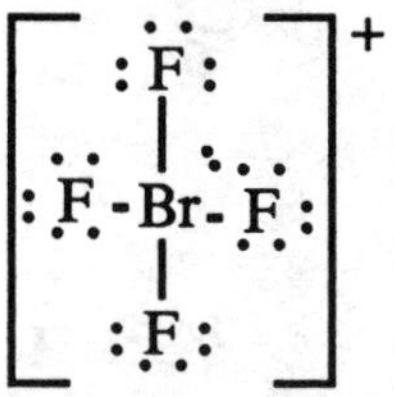

The ion has (5x7) - 1 = 34 valence electrons. The central
Br atom is surrounded by five electron groups--4 bonding
pairs and one lone pair: AX_4E. This corresponds to a
"seesaw" shape.

55. $:N \equiv N:$ An sp hybridization scheme for the N atoms is consistent with this
     Lewis structure. Alternatively, the σ bond could be described through the overlap
     of p_x orbitals and the π bonds through the parallel overlap of p_y orbitals and of p_z
     orbitals.

57. $\ddot{O}= C = C = C \overset{..}{=}\dot{O}$

According to VSEPR theory, the molecule should be linear,
and this corresponds to sp hybridization of all three C atoms.

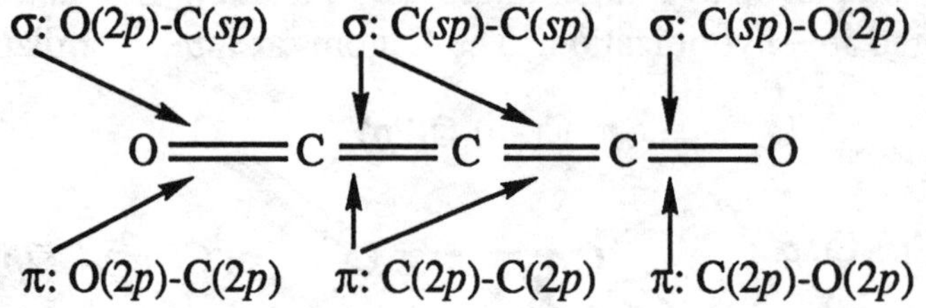

59. $\underline{\uparrow}$ σ_{2s}
     With one electron excited to a higher-energy

     $\underline{\uparrow}$ σ_{1s}^*
     bonding orbital the B.O. = 1. The electronically

     $\underline{\uparrow\downarrow}$ σ_{1s}
     excited He_2 molecule could exist.

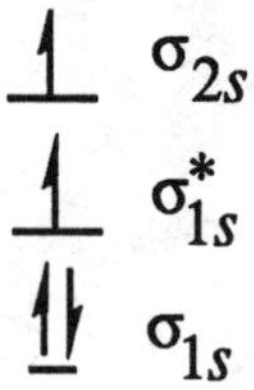

Chapter 11

Liquids, Solids, and Intermolecular Forces

Exercises

11.1 $\Delta H_{vap} = \frac{652 \text{ J}}{1.50 \text{ g } C_6H_6} \times \frac{78.11 \text{ g } C_6H_6}{\text{mol } C_6H_6} \times \frac{\text{kJ}}{10^3 \text{ J}} = 34.0 \text{ kJ/mol}$

11.2 $$m = \frac{PV\mathcal{M}}{RT} = \frac{19.8 \text{ mm } H_2O \times 275 \text{ mL} \times \frac{10^{-3} \text{ L}}{\text{mL}} \times 18.02 \text{ g/mol}}{\frac{62.4 \text{ L mmHg}}{\text{K mol}} \times (273 + 22)\text{K}}$$

$m = 5.33 \times 10^{-3} \text{ g } H_2O$

11.3 IBr has the greater molecular mass and so has the greater London forces. Both molecules are somewhat polar, but the electronegativity differences are small. IBr is a solid and BrCl is a gas.

11.4 NH_3, yes. H is bonded to N, a small electronegative atom.
CH_4, no. C is not electronegative enough for hydrogen bonding.
C_6H_5OH, yes. H is bonded to O, a small electronegative atom. Hydrogen bonding is somewhat overshadowed by dispersion forces associated with the large organic part of the molecule.
$CH_3\overset{\overset{\displaystyle O}{\|}}{C}OH$, yes. H is bonded to O, a small electronegative atom.
H_2S, no. There is a little hydrogen bonding because, although S is electronegative, S is too large to permit much hydrogen bonding.
H_2O_2, yes. H is bonded to O, a small electronegative atom.

11.5 $CsBr < KI < KCl < MgF_2$
A cesium ion is larger than a potassium ion. An iodide ion is larger than a chloride ion, which is larger than a fluoride ion. A magnesium ion is a small cation with a 2^+ charge, and exerts stronger interionic attractions than do either K^+ or Cs^+ ions.

11.6 There are 8 corner atoms, each unit cell has $\frac{1}{8}$ of each corner atom. There are 6 face centered atoms. Each unit cell has $\frac{1}{2}$ of each face centered atom.

corner atoms $\frac{1}{8} \times 8 = 1$

face atom $\frac{1}{2} \times 6 = 3$ $\quad 1 + 3 = 4$ atoms per unit cell

11.7 For each Na^+ ion in the center of the unit cell, there are 6 face centered Cl^- ions, making a coordination number of 6. For each Cl^- there are two Na^+ in the center of a cell and four on the edge of a cell. So the coordination number is again equal to 6.

	Na^+		Cl^-
center atom	$1 = 1$	corner atoms	$\frac{1}{8} \times 8 = 1$
edge atoms	$\frac{1}{4} \times 12 = \underline{3}$	face atoms	$\frac{1}{2} \times 6 = \underline{3}$
	4		4

The ratio 4 Na^+:4Cl^- corresponds to the 1:1 ratio in the formula NaCl.

Estimation Exercise

11.1 CS_2 $1.00 \text{ kg} \times \frac{10^3 \text{ g}}{\text{kg}} \times \frac{27.4 \text{ kJ}}{\text{mol}} \times \frac{\text{mol}}{76.13 \text{ g}} = 360 \text{ kJ}$

CCl_4 $1.00 \text{ kg} \times \frac{10^3 \text{ g}}{\text{kg}} \times \frac{37.0 \text{ kJ}}{\text{mol}} \times \frac{\text{mol}}{153.81 \text{ g}} = 240 \text{ kJ}$

CH_3OH $1.00 \text{ kg} \times \frac{10^3 \text{ g}}{\text{kg}} \times \frac{38.0 \text{ kJ}}{\text{mol}} \times \frac{\text{mol}}{32.04 \text{ g}} = 1.19 \times 10^3 \text{ kJ}$

C_8H_{18} $1.00 \text{ kg} \times \frac{10^3 \text{ g}}{\text{kg}} \times \frac{41.5 \text{ kJ}}{\text{mol}} \times \frac{\text{mol}}{115.52 \text{ g}} = 359 \text{ kJ}$

C_2H_5OH $1.00 \text{ kg} \times \frac{10^3 \text{ g}}{\text{kg}} \times \frac{43.3 \text{ kJ}}{\text{mol}} \times \frac{\text{mol}}{46.07 \text{ g}} = 940 \text{ kJ}$

H_2O $1.00 \text{ kg} \times \frac{10^3 \text{ g}}{\text{kg}} \times \frac{44.0 \text{ kJ}}{\text{mol}} \times \frac{\text{mol}}{18.02 \text{ g}} = 2.44 \times 10^3 \text{ kJ}$

$C_6H_5NH_2$ $1.00 \text{ kg} \times \frac{10^3 \text{ g}}{\text{kg}} \times \frac{52.3 \text{ kJ}}{\text{mol}} \times \frac{\text{mol}}{93.13 \text{ g}} = 562 \text{ kJ}$

A large ΔH value per mole and a small molar mass means H_2O requires the largest amount of heat. Conversely, CCl_4, with a relativley low ΔH value and a large molar mass requires the least.

11.2 a. CS_2 47 °C b. CH_3OH 65 °C c. CH_3CH_2OH 78 °C d. H_2O 100 °C

11.3 Hg $1.00 \text{ g} \times \frac{\text{mol}}{200.6 \text{ g}} \times \frac{2.30 \text{ kJ}}{\text{mol}} = 0.0115 \text{ kJ}$

CH_3CH_2OH $1.00 \text{ g} \times \frac{\text{mol}}{46.07 \text{ g}} \times \frac{5.01 \text{ kJ}}{\text{mol}} = 0.109 \text{ kJ}$

H_2O $1.00 \text{ g} \times \frac{\text{mol}}{18.02 \text{ g}} \times \frac{6.01 \text{ kJ}}{\text{mol}} = 0.334 \text{ kJ}$

C_6H_6 $1.00 \text{ g} \times \frac{\text{mol}}{78.11 \text{ g}} \times \frac{9.87 \text{ kJ}}{\text{mol}} = 0.126 \text{ kJ}$

Ag $1.00 \text{ g} \times \frac{\text{mol}}{107.9 \text{ g}} \times \frac{11.95 \text{ kJ}}{\text{mol}} = 0.111 \text{ kJ}$

Fe $1.00 \text{ g} \times \frac{\text{mol}}{55.85 \text{ g}} \times \frac{15.19 \text{ kJ}}{\text{mol}} = 0.272 \text{ kJ}$

Divide each ΔH_{fusion} by the molar mass and compare the values. Hg has the smallest ΔH and the largest molar mass; it requires the least heat. Water, with an intermediate value of ΔH_{fusion} (per mole) and the lowest molar mass, requires the most.

11.4 $\Delta H_{sub} = \Delta H_f + \Delta H_{vap}$

$$\Delta H_{vap} = \Delta H_{sub} - \Delta H_f = \frac{44.2\text{ kJ}}{\text{mol}} - \frac{9.87\text{ kJ}}{\text{mol}}$$

$\Delta H_{vap} = 34.3$ kJ/mol

Conceptual Exercise

11.1 Methane has a $T_c = -82.4$ °C at $P_c = 45$ atm. Room temperature is above T_c, so methane stays a gas at all pressures. The pressure gauge would work fine.

11.2 $$P = \frac{nRT}{V} = \frac{1.05\text{ mol} \times \frac{0.08206\text{ L atm}}{\text{K mol}} \times (273.2 + 30.0)\text{K}}{2.61\text{ L}}$$

$P = 10$ atm

This pressure greatly exceeds the vapor pressure of H_2O at 30.0 °C, so some of the vapor condenses to liquid. The sample cannot be all liquid, however, because 1.05 mol H_2O(l) occupies a volume of less than 20 mL. The final condition reached is one of liquid and vapor in equilibrium.

Review Questions

7. Each area represents one phase. The curves represent two phases in equilibrium. The triple point represents three phases in equilibrium.

12. a. Increased temperature increases the vapor pressure.
b., c. and d. Have no effect. The vapor pressure will be the same whatever the volume of liquid and vapor and the area of contact between them, as long as both phases exist together.

18. The polar substance will have a higher boiling point because it requires more energy to separate the polar molecules due to the presence of both dispersion forces and dipole-dipole forces.

19. CH_4 has only weak dispersion forces as intermolecular forces. H_2O has dispersion forces, dipole-dipole forces, and hydrogen bonds to hold the molecules together and in the liquid state.

20. Methane and ethane have only dispersion forces as intermolecular forces, whereas ethanol and methanol have dispersion, dipole-dipole, and hydrogen bonds to hold the molecules together.

21. Hexanoic acid has a higher molecular weight, but more important reasons are the dipole-dipole forces and hydrogen bonds that are present in hexanoic acid and absent in 2,2-dimethylbutane.

22. C_6H_5COOH has a high molecular weight, dipole-dipole forces, and hydrogen bonds; it should be a solid. $CH_3(CH_2)_8CH_3$ has only dispersion forces as intermolecular forces, and is likely to be a liquid. CH_3OH has dipole-dipole and hydrogen bonds, but a low molecular weight, and is also likely to be a liquid. $(CH_3CH_2)_2O$ has a higher molecular weight than CH_3OH but lacks hydrogen bonds; it is also likely to be a liquid.

23. BF_3 is the gas. It has a small molecular weight and is nonpolar.
NI_3 is a solid because of its large molecular weight and sufficiently large dispersion forces.
PCl_3 has a relatively high molecular weight compared to BF_3, but not as high as NI_3. PCl_3 is expected to be a liquid.
CH_3COOH has a molecular weight comparable to BF_3, but because of its polar character and ability to form hydrogen bonds, it is expected to be a liquid.

28. The larger the ionic charge and the smaller the ionic size, the larger is the lattice energy. The larger the lattice energy, the higher is the melting point.

Problems

31. Steam burns are often worse because the steam has more energy to transfer to the skin. It is at the same temperature as boiling water, but it transfers additional energy as the heat of condensation.

33. $\Delta H_T = \Delta H_1 + \Delta H_2$

25.0 °C <—> 15.5 °C ΔH_1 = mass x sp. ht. x ΔT

$$\Delta H_1 = 1.25 \text{ mol} \times \frac{32.04 \text{ g}}{\text{mol}} \times \frac{2.53 \text{ J}}{\text{g °C}} \times (15.5 - 25.0)\text{ °C} \times \frac{\text{kJ}}{10^3 \text{ J}}$$

$\Delta H_1 = -0.963$ kJ

vapor —> liquid $\Delta H_2 = -n\Delta H_{vap}$

$$\Delta H_2 = -1.25 \text{ mol} \times \frac{38.0 \text{ kJ}}{\text{mol}} = -47.5 \text{ kJ}$$

$\Delta H_T = -48.5$ kJ

35. a. about 430 mmHg
b. about 76 °C

37. The lowest boiling temperature for water is the triple point temperature (about 0 °C). The water has to be placed in a vacuum chamber and the pressure reduced until the water boils at that temperature. (Below this temperature the water would freeze.)

39. $n = \dfrac{PV}{RT}$

$$\frac{0.00143 \text{ mmHg} \times \dfrac{\text{atm}}{760 \text{ mmHg}} \times 27.5 \text{ m}^3 \times \left(\dfrac{\text{cm}}{10^{-2} \text{ m}}\right)^3 \times \dfrac{\text{mL}}{\text{cm}^3} \times \dfrac{10^{-3} \text{ L}}{\text{mL}}}{\dfrac{0.08206 \text{ L atm}}{\text{K mole}} \times (22 + 273)\text{K}} \times$$

$$\frac{6.022 \times 10^{23} \text{ atom}}{\text{mol}} = 1.29 \times 10^{21} \text{ atoms}$$

41. $$P = \frac{dRT}{\mathcal{M}} = \frac{\dfrac{0.876 \text{ g}}{\text{L}} \times \dfrac{62.4 \text{ L mmHg}}{\text{K mol}} \times (273 + 32)\text{K}}{58.08 \text{ g/mol}}$$

$P = 287$ mmHg

43. $P = \frac{mRT}{\mathcal{M}V} = \frac{0.625\text{ g} \times \frac{0.08206\text{ L atm}}{\text{K mol}} \times 273\text{ K}}{18.02\text{ g/mol} \times 178.5\text{ L}} = 0.00435\text{ atm}$

$0.00435\text{ atm} \times \frac{760\text{ mmHg}}{\text{atm}} = 3.31\text{ mmHg}$

The calculated pressure of $H_2O(g)$ is slightly less than the triple point pressure. No vapor condenses, either to solid or liquid. The system is one of gaseous water only.

45. $\Delta H = n\Delta H_{fus} = 3.55\text{ kg} \times \frac{10^3\text{ g}}{\text{kg}} \times \frac{\text{mol}}{18.02\text{ g}} \times \frac{6.01\text{ kJ}}{\text{mol}} = 1.18 \times 10^3\text{ kJ}$

47. $\Delta H = n(\Delta H_{fus} + \Delta H_{vap}) = n\Delta H_{sub}$

$\Delta H = 1.00\text{ kg} \times \frac{10^3\text{ g}}{\text{kg}} \times \frac{\text{mol}}{18.02\text{ g}} \times \left(\frac{44.0\text{ kJ}}{\text{mol}} + \frac{6.0\text{ kJ}}{\text{mol}}\right) = 2.77 \times 10^3\text{ kJ}$

49. $\Delta H_{ice} = -\Delta H_{H_2O}$

$n_{ice}\Delta H_{fus} + \text{mass}_{ice} \times \text{sp.ht.}_{ice} \times \Delta T_{ice} = -\text{mass}_{water} \times \text{sp.ht.}_{water} \times \Delta T_{water}$

$\left(25.5\text{ g} \times \frac{\text{mol}}{18.02\text{ g}} \times \frac{6.01\text{ kJ}}{\text{mol}}\right) + \left(25.5\text{ g} \times \frac{4.18\text{ J}}{\text{g °C}} \times (t - 0.0)\text{ °C}\right) \times \frac{\text{kJ}}{10^3\text{ J}} = -125\text{ mL} \times \frac{1.00\text{ g}}{\text{mL}} \times \frac{4.18\text{ J}}{\text{g °C}} \times (t - 26.5)\text{ °C} \times \frac{\text{kJ}}{10^3\text{ J}}$

$8.50\text{ kJ} + 0.107\frac{t\text{kJ}}{\text{°C}} = -0.523\frac{t\text{kJ}}{\text{°C}} + 13.8\text{ kJ}$

$0.630\frac{t\text{kJ}}{\text{°C}} = 5.3\text{ kJ}$

$t = 8.4\text{ °C}.$

51. The enthalpy change should be the same for the two step process as for direct sublimation. Enthalpy is a state function so is path independent.

53.

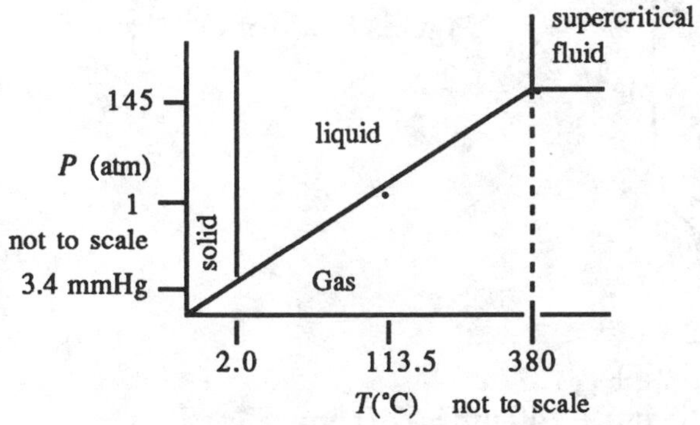

55. The solid CO_2 becomes a liquid at a temperature somewhat above -56.7 °C (the triple point is 5.1 atm and -56.7 °C). The liquid then converts to gaseous CO_2 at a temperature that is probably below room temperature (the critical point is at 72.9 atm and 304.2 K).

57.   Even though PF_3 is somewhat more polar than PI_3, strong dispersion forces in the high-molecular-weight PI_3 cause it to persist as a liquid at much higher temperatures than PF_3. PF_3 has the lower boiling point.

59.   2,2-Dimethyl-1-butanol has a higher molecular weight so more dispersion forces, but more importantly, it can also form hydrogen bonds. It has the higher boiling point.

61.   The substances have similar molecular weights, but the compactness of the 3,3-dimethylpentane molecule and the possibility of hydrogen bonding in 1-pentanol account for 1-pentanol having the higher melting point.

63. $CH_4 < CH_3CH_3 < NH_3 < H_2O$
        -161   -89   H bond H bond
                             -33   100
        CH_4 and CH_3CH_3 are nonpolar. NH_3 has a higher boiling point than CH_3CH_3, despite ethane's higher molecular weight, because of hydrogen bonding. In H_2O, hydrogen bonding is stronger still.

65. $CH_3OH < C_6H_5OH < NaOH < LiOH$
        H bond   H bond   ionic   ionic
        -97.8   40.9   318   450
        C_6H_5OH has a higher melting point than CH_3OH because of its greater molecular weight. The ionic compounds have higher melting points than the covalent compounds, and that of LiOH is greater than NaOH because of the smaller size of the Li^+ ion.

67.   Ethylene glycol can form more hydrogen bonds than isopropyl alcohol, so ethylene glycol has stronger intermolecular forces and a higher surface tension.

69.   Sodium chloride is FCC for the Cl^-. In Figure 11.33, replace Na^+ by Mg^{2+} and Cl^- by O^{2-}.

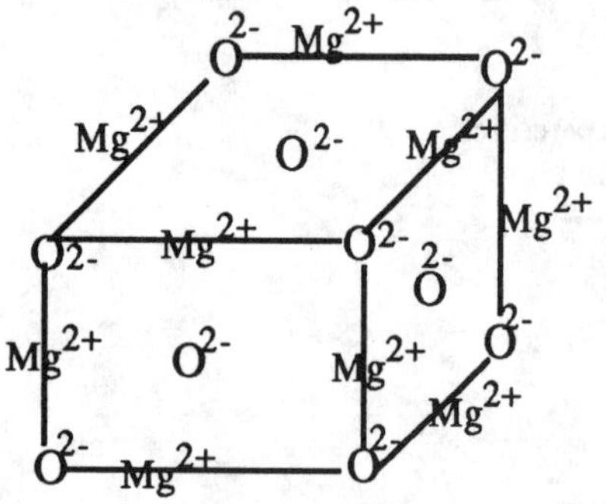

MgO is FCC for O^{2-}

length $= r_{O^{2-}} + 2 \times r_{Mg^{2+}} + r_{O^{2-}}$
       $= 140 \text{ pm} + 2 \times 65 \text{ pm} + 140 \text{ pm} = 410 \text{ pm}$

71.   The surface tension of water must be reduced so that the water molecules will have less affinity for each other and a greater affinity for the surface.

73. $\frac{P_1V_1}{T_1} = \frac{P_2V_2}{T_2}$, $P_2 = \frac{P_1V_1}{V_2} = \frac{750.0 \text{ mmHg x } 150.0\text{mL}}{172 \text{ mL}}$

For N_2 gas $P_2 = 654$ mmHg

$P_{total} = P_{N_2} + P_{C_6H_6}$

$P_{C_6H_6} = 750.0 \text{ mmHg} - 654 \text{ mmHg} = 96 \text{ mmHg}$

75. a. $\log P = 6.876 - \frac{1171}{25 + 224.4} = 2.181$

$10^{\log P} = P = 10^{2.181} = 152 \text{ mmHg}$

b. $\log(760) = 6.876 - \frac{1171}{t + 224.4}$

$2.881 = 6.876 - \frac{1171}{t + 224.4}$

$3.995 = \frac{1171}{t + 224.4}$

$3.995\,t + 896 = 1171$

$3.995\,t = 275$

$t = 68.8\ °\text{C}$

77. $2 H_2 (g) + O_2 (g) \longrightarrow 2 H_2O (l)$

$1.00 \text{ g } H_2 \text{ x } \frac{\text{mol}}{2.016 \text{ g}} \text{ x } \frac{2 \text{ mol } H_2O}{2 \text{ mol } H_2} = 0.4960 \text{ mol } H_2O$

$1.00 \text{ g } H_2 \text{ x } \frac{\text{mol}}{2.016 \text{ g}} \text{ x } \frac{\text{mol } O_2}{2 \text{ mol } H_2} = 0.2480 \text{ mol } O_2 \text{ used}$

$10.00 \text{ g x } \frac{\text{mol}}{32.00 \text{ g}} = 0.3125 \text{ mol } O_2 \text{ original}$

$0.3125 \text{ mol} - 0.2480 \text{ mol} = 0.0645 \text{ mol excess } O_2$

$$P_{O_2} = \frac{nRT}{V} = \frac{0.0645 \text{ moles x } \frac{0.08206 \text{ L atm}}{\text{K mol}} \text{ x } (25 + 273)\text{K}}{3.15 \text{ L}}$$

$P_{O_2} = 0.501 \text{ atm} = 381 \text{ mmHg}$

The *V.P.* of H_2O at 25 °C is 23.8 mmHg.

$P_{total} = P_{O_2} + P_{H_2O}$

$P_{total} = 381 \text{ mmHg} + 23.8 \text{ mmHg}$

$P_{total} = 405 \text{ mmHg}$

79. $$P = \frac{mRT}{\mathcal{M}V} = \frac{1.00 \text{ g x } \frac{0.08206 \text{ L atm}}{\text{K mol}} \text{ x } (35 + 273)\text{K}}{18.02 \text{ g/mol x } 40.0\text{L}}$$

$P = 0.03508 \text{ atm x } \frac{760 \text{ mm Hg}}{\text{atm}} = 26.66 \text{ mm Hg}$

At 27 °C the pressure in the chamber will be equal to the vapor pressure of water and liquid will begin to form.

81. $$P = \frac{mRT}{\mathcal{M}V} = \frac{1510\text{ g} \times \dfrac{0.08206\text{ L atm}}{\text{K mol}} \times (25 + 273)\text{K}}{44.01\text{ g/mol} \times 11.5\text{ L}}$$

$P = 73$ atm

Because this pressure exceeds the vapor pressure of CO_2 (l) at 25 °C, liquid and vapor are present together. It is unlikely that the CO_2 could exist as liquid only, for if it did, its density would be only d = 1510 g/11500mL = 0.131 g/mL. (The actual density of CO_2(l) is more nearly about 1 g/mL.)

83.

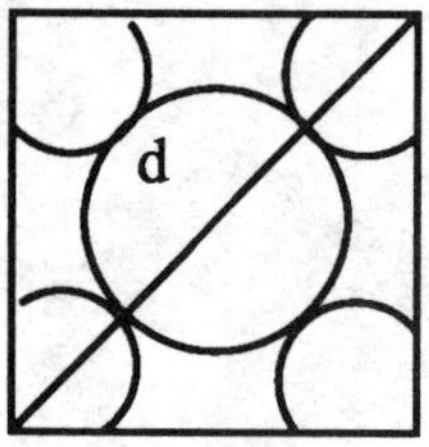

Three atoms are in contact along the diagonal. The length of the diagonal = $4r$.

Pythagorean formula $a^2 + b^2 = d^2$

For unit cell $a = b = l$ length of cell

$$l^2 + l^2 = 2l^2 = d^2 = (4r)^2$$

$$2l^2 = 16r^2$$

$$l = \frac{4r}{\sqrt{2}} = \frac{4 \times 143\text{ pm}}{\sqrt{2}} = 404\text{ pm}$$

$$V = l^3 = (404\text{ pm})^3 \frac{(10^{-10}\text{cm})^3}{\text{pm}^3} = 6.59 \times 10^{-23}\text{cm}^3$$

85. In simple cubic, there is the equivalent of one atom in a unit cell of length $2r$.

$$V_{cell} = (2r)^3 = 8r^3$$

$$V_{atom} = 4/3\ \pi r3 = 4.1888\ r^3$$

$$\text{void \%} = \frac{(8.000 - 4.1888)\ r^3}{8\ r^3} \times 100\% = 47.64\%$$

In FCC there is the equivalent of 4 atoms. As shown in Problem 83, the volume of the cell is $V_{cell} = l^3 = \left(\frac{4r}{\sqrt{2}}\right)^3 = 22.627\ r^3$

$$V_{atoms} = 4 \times \frac{4}{3}\pi r^3 = 16.755\ r^3$$

$$\text{void \%} = \frac{22.627\ r^3 - 16.755\ r^3}{22.627\ r^3} \times 100\% = 25.95\%$$

Chapter 12

Chemical Reactions in Aqueous Solutions

Exercises

12.1 $[\text{glucose}] = \dfrac{20.0\ \text{g}}{\text{L}} \times \dfrac{\text{mol}}{180.2\ \text{g}} = 0.111\ \text{M glucose}$

Molarity of sodium citrate $= \dfrac{2.9\ \text{g}}{\text{L}} \times \dfrac{\text{mol}}{258.1\ \text{g}} = 0.0112\ \text{M sodium citrate}$

Molarity of KCl $= \dfrac{1.5\ \text{g}}{\text{L}} \times \dfrac{\text{mol}}{74.55\ \text{g}} = 0.0201\ \text{M KCl}$

Molarity of NaCl $= \dfrac{3.5\ \text{g}}{\text{L}} \times \dfrac{\text{mol}}{58.44\ \text{g}} = 0.0599\ \text{M NaCl}$

$[\text{K}^+] = 0.0201\ \text{M}$
$[\text{Cl}^-] = 0.0201\ \text{M} + 0.0599\ \text{M} = 0.0800\ \text{M}$
$[\text{Na}^+] = 0.0599\ \text{M} + 3 \times 0.0112\ \text{M} = 0.0935\ \text{M}$

12.2 a. $Ca(OH)_2\ (s) + 2\ HCl\ (aq) \rightarrow CaCl_2\ (aq) + 2H_2O\ (l)$
b. $Ca^{2+}\ (aq) + 2\ OH^-\ (aq) + 2\ H^+\ (aq) + 2\ Cl^-\ (aq) \rightarrow Ca^{2+}\ (aq) + 2\ Cl^-\ (aq)$
$+ 2\ H_2O\ (l)$

c. $OH^-\ (aq) + H^+\ (aq) \rightarrow H_2O\ (l)$

12.3 $10.00\ \text{mL} \times \dfrac{4.12\ \text{g HC}_2\text{H}_3\text{O}_2}{100\ \text{g solution}} \times \dfrac{1.01\ \text{g}}{\text{mL}} \times \dfrac{\text{mL}}{10^{-3}\ \text{L}} \times \dfrac{\text{mol HC}_2\text{H}_3\text{O}_2}{60.05\ \text{g}}$

$\times \dfrac{\text{mol OH}^-}{\text{mol HC}_2\text{H}_3\text{O}_2} \times \dfrac{1\ \text{L}}{0.550\ \text{mol OH}} = 12.6\ \text{mL NaOH}$

12.4 a. $MgSO_4\ (aq) + 2\ KOH\ (aq) \rightarrow Mg(OH)_2\ (s) + 2\ K^+\ (aq) + SO_4^{2-}\ (aq)$
$Mg^{2+}\ (aq) + OH^-\ (aq) \rightarrow Mg(OH)_2\ (s)$
b. $2\ FeCl_3\ (aq) + 3\ Na_2S\ (aq) \rightarrow Fe_2S_3\ (s) + 6\ Na^+\ (aq) + 6\ Cl^-\ (aq)$
$2\ Fe^{3+}\ (aq) + 3\ S^{2-}\ (aq) \rightarrow Fe_2S_3\ (s)$
c. $CaCl_2$ and Na_2CO_3 are soluble, and $CaCO_3$ (s) is insoluble; there is no reaction.

12.5 a. $\overset{+3\ -2}{Al_2O_3}$ b. $\overset{0}{P_4}$ c. $\overset{+1\ +7\ -2}{NaMnO_4}$ d. $\overset{+1\ -2}{ClO^-}$

e. $\overset{+1\ +5\ -2}{HAsO_4^{2-}}$ f. $\overset{+1\ +5\ -1}{HSbF_6}$ g. $\overset{+1\ -\frac{1}{2}}{CsO_2}$ h. $\overset{-2\ +1\ -1}{CH_3F}$

i. $\overset{+2\ +1\ -1}{CHCl_3}$ j. $\overset{0\ +1\ \ 0\ -2\ -2\ +1}{CH_3COOH}$

12.6 $\overset{0}{I_2} + \overset{0}{Cl_2} + H_2O \rightarrow \overset{+5}{IO_3^-} + H^+ + \overset{-1}{Cl^-}$

Yes, I_2 is oxidized and Cl_2 is reduced.

12.7 Write two skeleton half-equations.
$MnO_4^- \rightarrow Mn^{2+}$
$C_2O_4^{2-} \rightarrow CO_2$
Balance all atoms except H and O.
$C_2O_4^{2-} \rightarrow 2\ CO_2$
Balance O by adding H_2O.
$MnO_4^- \rightarrow Mn^{2+} + 4\ H_2O$
Balance H by adding H^+.
$MnO_4^- + 8\ H^+ \rightarrow Mn^{2+} + 4\ H_2O$
Balance charge by adding e^-s.
$MnO_4^- + 8\ H^+ + 5\ e^- \rightarrow Mn^{2+} + 4\ H_2O$
$C_2O_4^{2-} \rightarrow 2\ CO_2 + 2\ e^-$
Multiply to eliminate e^-s and add the half-equations.
$(MnO_4^- + 8\ H^+ + 5\ e^- \rightarrow Mn^{2+} + 4\ H_2O) \times 2$
$(C_2O_4^{2-} \rightarrow 2\ CO_2 + 2\ e^-) \times 5$

$2\ MnO_4^- + 5\ C_2O_4^{2-} + 16\ H^+ \rightarrow 2\ Mn^{2+} + 10\ CO_2 + 8\ H_2O$

12.8 Write the two skeleton half-equations.
$2\ CN^- \rightarrow 2\ CO_3^{2-} + N_2$
$OCl^- \rightarrow Cl^-$
Balance O by adding H_2O.
$2\ CN^- + 6\ H_2O \rightarrow 2\ CO_3^{2-} + N_2$
$OCl^- \rightarrow Cl^- + H_2O$
Balance H by adding H^+.
$2\ CN^- + 6\ H_2O \rightarrow 2\ CO_3^{2-} + N_2 + 12\ H^+$
$OCl^- + 2\ H^+ \rightarrow Cl^- + H_2O$
Balance charge by adding e^-s.
$2\ CN^- + 6\ H_2O \rightarrow 2\ CO_3^{2-} + N_2 + 12\ H^+ + 10\ e^-$
$OCl^- + 2\ H^+ + 2\ e^- \rightarrow Cl^- + H_2O$
Multiply to eliminate e^-s and add the half-reactions.
$2\ CN^- + 6\ H_2O \rightarrow 2\ CO_3^{2-} + N_2 + 12\ H^+ + 10\ e^-$
$(OCl^- + 2\ H^+ + 2\ e^- \rightarrow Cl^- + H_2O) \times 5$

$2\ CN^- + 5\ OCl^- + H_2O \rightarrow 2\ CO_3^{2-} + N_2 + 5\ Cl^- + 2\ H^+$
Add 2 OH^- to both sides corresponding to 2 H^+ on right) and subtract one H_2O from both sides.
$2\ CN^- + 5\ OCl^- + H_2O + 2\ OH^- \rightarrow 2\ CO_3^{2-} + N_2 + 5\ Cl^- + 2\ H^+ + 2\ OH^-$
$2\ CN^- + 5\ OCl^- + 2\ OH^- \rightarrow 2\ CO_3^{2-} + N_2 + 5\ Cl^- + H_2O$

12.9 $\overset{+3}{Fe}(OH)_3 + O\overset{+1}{Cl}^- + OH^- \rightarrow \overset{+6}{Fe}O_4^{2-} + \overset{-1}{Cl}^- + H_2O$

-3e (Fe), +2e$^-$ (Cl)

$2\ Fe(OH)_3 + 3\ OCl^- +$ ___ $OH^- \rightarrow 2\ FeO_4^{2-} + 3\ Cl^- +$ ___ H_2O

balance charge
$2\ Fe(OH)_3 + 3\ OCl^- + 4\ OH^- \rightarrow 2\ FeO_4^{2-} + 3\ Cl^- +$ ___ H_2O

balance H and O atoms
$2\ Fe(OH)_3 + 3\ OCl^- + 4\ OH^- \rightarrow 2\ FeO_4^{2-} + 3\ Cl^- + 5\ H_2O$

12.10 $0.2865\ \text{g sample} \times \frac{58.01\ \text{g Fe}}{100\ \text{g sample}} \times \frac{\text{mol } Fe^{2+}}{55.847\ \text{g } Fe^{2+}}$

$\times \frac{1\ \text{mol } Cr_2O_7^{2-}}{6\ \text{mol } Fe^{2+}} \times \frac{1}{0.02250\ \text{M } Cr_2O_7^{2-}} \times \frac{\text{mL}}{10^{-3}\ \text{L}} = 22.04\ \text{mL}$

Alternatively, because the $K_2Cr_2O_7$ solution in Exercise 12.10 has the same molarity as the $KMnO_4$ of Example 12.10 and the stoichiometric factors are 1 mol $Cr_2O_7^{2-}$/ 6 mol Fe^{2+} and 1 mol MnO_4^- / 5 mol Fe^{2+}, the volume of $K_2Cr_2O_7$ (aq) required is 5/6 of that of $KMnO_4$ (aq).
5/6 x 26.45 mL = 22.04 mL

Conceptual Exercises

12.1 The CH_3NH_2 will cause a dimly lit bulb, as it is a weak base. HNO_3 is a strong acid and will cause a brightly lit bulb. The combination will produce a strong electrolyte and a brightly lit bulb.

$CH_3NH_2 + H_2O \rightleftharpoons CH_3NH_3^+ + OH^-$
$HNO_3 \rightarrow H^+ + NO_3^-$
$H^+ + OH^- \rightarrow H_2O$

$CH_3NH_2 + HNO_3 \rightarrow CH_3NH_3^+ + NO_3^-$

12.2 The acid will cause the $Fe(OH)_3$ to dissolve because the H^+ will react with the OH^- to form H_2O.

$Fe(OH)_3\ (s) + 3\ H^+\ (aq) \rightarrow Fe^{3+}\ (aq) + 3\ H_2O\ (l)$

12.3 $Cr_2O_7^{2-}$ (aq) is an oxidizing agent; it will react with a reducing agent. HNO_3 (aq) is also an oxidizing agent. There is no reaction between $Cr_2O_7^{2-}$ (aq) and HNO_3 (aq). HCl is a reducing agent (with Cl in the oxidation state, -1). It is oxidized by $Cr_2O_7^{2-}$ (aq), probably to Cl_2 (g).

Review Questions

1. non-electrolytes: (a), (h)
strong electrolytes: (b), (c), (e), (g)
weak electrolytes: (d), (f)

2. a. salt b. strong base c. salt d. weak acid e. strong acid
f. weak base g. salt h. strong base

3. (c) is highest, with $[NO_3^-] = 3 \times 0.040 = 0.12$ M.
(a) and (d) both have $[NO_3^-] = 0.10$ M.
(b) is lowest, with $[NO_3^-] = 0.080$ M

4. (a) $[Al^{3+}] = 0.0036$ M and (c) $[Al^{3+}] = 0.0040$ M

5. (d) [ions] = 0.075 M

6. 0.10 M NaCl has more ions in solution. It is the only strong electrolyte.

7. 0.10 M H_2SO_4. It produces $[H^+] > 0.10$ M, because ionization is complete in the first ionization step and also occurs to some extent in the second.

10. $Ca(OH)_2\ (s) + 2\ H^+\ (aq) \longrightarrow Ca^{2+}\ (aq) + 2\ H_2O$

11. $CaCO_3\ (s) + 2\ CH_3COOH \longrightarrow Ca^{2+} + 2\ CH_3COO^- + H_2CO_3$
$\quad \hookrightarrow H_2O + CO_2\ (g)$

12. $HCO_3^-\ (aq) + HCOOH\ (aq) \longrightarrow HCOO^-\ (aq) + H_2CO_3$
$\quad \hookrightarrow H_2O + CO_2\ (g)$

13. The solution contains one or a combination of the ions Ag^+, Pb^{2+}, Hg_2^{2+}.

15. Na_2CO_3. Most carbonates are insoluble.

19. $Cu^{2+}\ (aq) + 2\ OH^-\ (aq) \longrightarrow Cu(OH)_2\ (s)$

20. $2Fe^{3+}\ (aq) + 3S^{2-}\ (aq) \longrightarrow Fe_2S_3\ (s)$

21. H is usually +1, O is usually -2. Hydrides have H as -1. Peroxides have O as -1, superoxides as -1/2.

25.

0	+3	-2	+6	-3
a. Cr	b. ClO_2^-	c. K_2Se	d. TeF_6	e. PH_4^+
+4	+4	+5	+2.5	-1
f. $CaRuO_3$	g. $SrTiO_3$	h. $P_2O_7^{4-}$	i. $S_4O_6^{2-}$	j. NH_2OH

26.

a. -3	b. +2	c. 0 0
C_2H_6	HCOOH	CH_3CHO
d. -2	e. -2 -2 -2	
$(CH_3)_2O$	$CH_3(CH_2)_3CH_2OH$	

Problems

29. a. $[Li^+] = 0.0385$ M b. $[Cl^-] = 0.070$ M
c. $[Al^{3+}] = 0.0224$ M d. $[Na^+] = 0.24$ M

31. $[NO_3^-] = \frac{0.112\text{ g}}{125\text{ mL}} \times \frac{\text{mL}}{10^{-3}\text{ L}} \times \frac{\text{mol Mg(NO}_3)_2\cdot 6\text{ H}_2\text{O}}{256.4\text{ g}} \times \frac{2\text{ mol NO}_3^-}{\text{mol Mg(NO}_3)_2\cdot 6\text{ H}_2\text{O}} = 6.99$
$\times 10^{-3}$ M

33. $[Na^+] = 0.0554 + (2 \times 0.0145) = 0.0844$ M,
$[Cl^-] = 0.0554$ M, $[SO_4^{2-}] = 0.0145$ M

35. $V = 0.250\text{ L} \times \frac{0.0135\text{ mol Cl}^-}{\text{L}} \times \frac{1\text{ mol MgCl}_2}{2\text{ mol Cl}^-} \times \frac{1\text{ L}}{0.0250\text{ M MgCl}_2} \times \frac{\text{mL}}{10^{-3}}$
$= 67.5$ mL

37. a. $HI\ (aq) \longrightarrow H^+\ (aq) + I^-\ (aq)$
b. $CH_3CH_2COOH\ (aq) \leftrightarrow CH_3CH_2COO^-\ (aq) + H^+\ (aq)$
c. $HNO_2\ (aq) \Leftrightarrow H^+\ (aq) + NO_2^-\ (aq)$
d. $H_2PO_4^-\ (aq) \Leftrightarrow H^+\ (aq) + HPO_4^{2-}$

39. a. $25.00\text{ mL} \times \frac{10^{-3}\text{ L}}{\text{mL}} \times 0.0365\text{ M KOH} \times \frac{\text{mol HCl}}{\text{mol KOH}} \times \frac{1}{0.0195\text{ M}} \times \frac{\text{mL}}{10^{-3}\text{ L}}$
$= 46.8$ mL

b. $10.00\text{ mL} \times \frac{10^{-3}\text{ L}}{\text{mL}} \times 0.0116\text{ M Ca(OH)}_2 \times \frac{2\text{ mol HCl}}{\text{mol Ca(OH)}_2} \times \frac{1}{0.0195\text{ M}} \times \frac{\text{mL}}{10^{-3}\text{ L}}$
$= 11.9$ mL

c. $20.00\text{ mL} \times \frac{10^{-3}\text{ L}}{\text{mL}} \times 0.0225\text{ M NH}_3 \times \frac{\text{mol HCl}}{\text{mol NH}_3} \times \frac{1}{0.0195\text{ M}} \times \frac{\text{mL}}{10^{-3}\text{ L}}$
$= 23.1$ mL

41. $CH_3COOH + OH^- \longrightarrow H_2O + CH_3COO^-$
$31.45\text{ mL} \times \frac{10^{-3}\text{ L}}{\text{mL}} \times 0.2560\text{ M KOH} \times \frac{\text{mol CH}_3\text{COOH}}{\text{mol KOH}} \times \frac{1}{10.00\text{ mL}} \times \frac{\text{mL}}{10^{-3}\text{ L}}$
$= 0.8051$ M

43. $148\text{ kg Na}_2\text{CO}_3 \times \frac{10^3\text{ g}}{\text{kg}} \times \frac{\text{mol}}{105.99\text{ g}} \times \frac{\text{mol CO}_2}{\text{mol Na}_2\text{CO}_3} \times \frac{0.08206\text{ L atm}}{\text{K mol}} \times$

$$\frac{(22 + 273)\text{K}}{748\text{ mmHg} \times \frac{\text{atm}}{760\text{ mmHg}}} = 3.43 \times 10^4\text{ L CO}_2$$

45. $RbOH\ (aq) + HCl\ (aq) \longrightarrow RbCl\ (aq) + H_2O\ (l)$
$Rb^+\ (aq) + OH^-\ (aq) + H^+\ (aq) + Cl^-\ (aq) \longrightarrow Rb^+\ (aq) + Cl^-\ (aq) + H_2O\ (l)$
$OH^-\ (aq) + H^+\ (aq) \longrightarrow H_2O\ (l)$

47. a. $2\,I^- + Pb^{2+} \longrightarrow PbI_2\ (s)$
b. No reaction
c. $Cr^{3+} + 3\,OH^- \longrightarrow Cr(OH)_3\ (s)$
d. No reaction
e. $OH^- + H^+ \longrightarrow H_2O\ (l)$
f. $HSO_4^- + OH^- \longrightarrow H_2O\ (l) + SO_4^{2-}$

49. a. $MgO\ (s) + 2\,H^+\ (aq) \longrightarrow Mg^{+2}\ (aq) + H_2O\ (l)$
b. $HCOOH\ (aq) + NH_3\ (aq) \longrightarrow NH_4^+\ (aq) + HCOO^-\ (aq)$
c. No reaction
d. $Cu^{2+}\ (aq) + CO_3^{2+}\ (aq) \longrightarrow CuCO_3\ (s)$
e. No reaction

51. a. $ClO_2 + H_2O \longrightarrow ClO_3^- + 2\,H^+ + e^-$ oxidation
b. $MnO_4^- + 4\,H^+ + 3\,e^- \longrightarrow MnO_2 + 2\,H_2O$ reduction
c. $2\,BrO^- + 4\,H^+ + 2\,e^- \longrightarrow Br_2 + 2\,H_2O$
$2\,BrO^- + 4\,H^+ + 4\,OH^- + 2\,e^- \longrightarrow Br_2 + 2\,H_2O + 4\,OH^-$
$2\,BrO^- + 2\,H_2O + 2\,e^- \longrightarrow Br_2 + 4\,OH^-$ reduction
d. $SbH_3 \longrightarrow Sb + 3\,H^+ + 3\,e^-$
$SbH_3 + 3\,OH^- \longrightarrow Sb + 3\,H^+ + 3\,OH^- + 3\,e^-$
$SbH_3 + 3\,OH^- \longrightarrow Sb + 3\,H_2O + 3\,e^-$ oxidation

53. The solution to the balancing of redox equations is listed in this order: first, the two balanced half-reactions are listed side by side. These are followed by the final balanced equation for acidic solution. For basic solutions, two more equations are listed. One shows the addition of the OH^- ions. Finally is listed the balanced equation in basic solution.
a. $Ag \longrightarrow Ag^+ + e^-$ $\quad NO_3^- + 4\,H^+ + 3\,e^- \longrightarrow NO\ (g) + 2\,H_2O$
$3\,Ag + NO_3^- + 4\,H^+ \longrightarrow 3\,Ag^+ + NO\ (g) + 2\,H_2O$
b. $H_2O_2 \longrightarrow O_2 + 2\,H^+ + 2\,e^-$ $\quad MnO_4^- + 8\,H^+ + 5\,e^- \longrightarrow Mn^{2+} + 4\,H_2O$
$2\,MnO_4^- + 5\,H_2O_2 + 6\,H^+ \longrightarrow 5\,O_2 + 2\,Mn^{2+} + 8\,H_2O$
c. $PbO + H_2O \longrightarrow PbO_2 + 2\,H^+ + 2\,e^-$ $\quad V^{3+} + H_2O \longrightarrow VO^{2+} + 2\,H^+ + e^-$
As both PbO_2 and V^{3+} are being oxidized, this is not possible.

55. a. $CN^- + H_2O \longrightarrow OCN^- + 2\,H^+ + 2\,e^-$ $\quad BrO_3^- + 6\,H^+ + 6\,e^- \longrightarrow Br^- + 3\,H_2O$
$3\,CN^- + BrO_3^- \longrightarrow 3\,OCN^- + Br^-$
b. $S_8 + 12\,H_2O \longrightarrow 4\,S_2O_3^{2-} + 24\,H^+ + 16\,e^-$ $\quad S_8 + 16\,e^- \longrightarrow 8\,S^{2-}$
$2\,S_8 + 12\,H_2O \longrightarrow 4\,S_2O_3^{2-} + 8\,S^{2-} + 24\,H^+$
$2\,S_8 + 12\,H_2O + 24\,OH^- \longrightarrow 4\,S_2O_3^{2-} + 8\,S^{2-} + 24\,H^+ + 24\,OH^-$
$S_8 + 12\,OH^- \longrightarrow 2\,S_2O_3^{2-} + 4\,S^{2-} + 6\,H_2O$

57. a. $NO + 5\,H^+ + 5\,e^- \longrightarrow NH_3 + H_2O$ $\quad H_2 \longrightarrow 2\,H^+ + 2\,e^-$
$2\,NO + 5\,H_2 \longrightarrow 2\,NH_3 + 2\,H_2O$
b. $8\,Fe_2S_3 + 48\,H_2O \longrightarrow 16\,Fe(OH)_3 + 3\,S_8 + 48\,H^+ + 48\,e^-$
$O_2 + 4\,H^+ + 4\,e^- \longrightarrow 2\,H_2O$
$8\,Fe_2S_3 + 12\,O_2 + 24\,H_2O \longrightarrow 16\,Fe(OH)_3 + 3\,S_8$

59. a. $MnO_4^- + 8\,H^+ + 5\,e^- \longrightarrow Mn^{2+} + 4\,H_2O$ $\quad C_2H_2O_4 \longrightarrow 2\,CO_2 + 2\,H^+ + 2\,e^-$

$2\,MnO_4^- + 5\,C_2H_2O_4 + 6\,H^+ \longrightarrow 2\,Mn^{2+} + 10\,CO_2 + 8\,H_2O$

b. $MnO_4^- + 4H^+ + 3\,e^- \longrightarrow MnO_2 + 2\,H_2O$

$C_2H_4O + H_2O \longrightarrow C_2H_4O_2 + 2\,H^+ + 2\,e^-$

$2\,MnO_4^- + 3\,C_2H_4O + 2\,H^+ \longrightarrow 2\,MnO_2 + 3\,C_2H_4O_2 + H_2O$

$2\,MnO_4^- + 3\,C_2H_4O + 2\,H^+ + 2\,OH^- \longrightarrow$

$2\,MnO_2 + 3\,C_2H_4O_2 + H_2O + 2\,OH^-$

$2\,MnO_4^- + 3\,C_2H_4O + H_2O \longrightarrow 2\,MnO_2 + 3\,C_2H_4O_2 + 2\,OH^-$

61.

	oxidizing agent	reducing agent
53a	NO_3^-	Ag
b	MnO_4^-	H_2O_2
c	--	--
54a	ClO_3^-	Mn^{2+}
b	--	--
c	O_2	S_8

63. $Mn^{2+} + 2\,H_2O \rightarrow MnO_2 + 4\,H^+ + 2\,e^-$

$MnO_4^- + 4\,H^+ + 3\,e^- \rightarrow MnO_2 + 2\,H_2O$

$3\,Mn^{2+} + 2\,MnO_4^- + 2\,H_2O \rightarrow 5\,MnO_2 + 4\,H^+$

$3\,Mn^{2+} + 2\,MnO_4^- + 4\,OH^- \rightarrow 5\,MnO_2 + 2\,H_2O$

$$0.03477\ \text{L} \times 0.05876\ \text{M}\ MnO_4^- \times \frac{3\ \text{mol}\ Mn^{2+}}{2\ \text{mol}\ MnO_4^-} \times \frac{1}{0.02500\ \text{L}} = 0.1226\ \text{M}\ Mn^{2+}$$

65. $C_2O_4^{2-} \rightarrow 2\,CO_2 + 2\,e^-$ $\quad MnO_4^- + 8\,H^+ + 5\,e^- \rightarrow Mn^{2+} + 4\,H_2O$

$2\,MnO_4^- + 5\,C_2O_4^{2-} + 16\,H^+ \rightarrow 10\,CO_2 + 2\,Mn^{2+} + 8\,H_2O$

$$0.02140\ \text{M}\ KMnO_4 \times \frac{10^{-3}\ \text{L}}{\text{mL}} \times 25.82\ \text{mL} \times \frac{5\ \text{mol}\ C_2O_4^{2-}}{2\ \text{mol}\ MnO_4^-}$$

$$\times \frac{250\ \text{mL}}{5.00\ \text{mL}} \times \frac{134.0\ \text{g}\ Na_2C_2O_4}{\text{mol}\ Na_2C_2O_4} = 9.26\ \text{g}\ Na_2C_2O_4$$

67. NH_3 and $Ba(OH)_2$ are bases. CH_3CH_2COOH is a weak acid. HI is a strong acid, as is H_2SO_4 in its first ionization. Because H_2SO_4 ionizes further in a second step, it produces the highest $[H^+]$.

69. $$[K^+] = \frac{2.46\ \text{mg}}{10.5\ \text{L}} \times \frac{10^{-3}\ \text{g}}{\text{mg}} \times \frac{\text{mol}\ KNO_3}{101.11\ \text{g}} \times \frac{1\text{mol}\ K^+}{1\text{mol}\ KNO_3} = 2.32 \times 10^{-6}\ \text{M}$$

$$[K^+] = \frac{45.5\ \text{g}\ K^+}{10^6\ \text{g}\ H_2O} \times \frac{1.00\text{g}}{\text{mL}} \times \frac{\text{mL}}{10^{-3}\ \text{L}} \times \frac{\text{mol}\ K^+}{39.10\ \text{g}\ K^+} = 1.16 \times 10^{-3}\ \text{M}$$

The solution with 45.5 ppm K^+ has the higher molarity of K^+.

71. $[I^-] = 0.0240\ \text{M} + 2 \times 0.0146\ \text{M} = 0.0532\ \text{M}$

$$V = 0.1000\ \text{L} \times \frac{0.0532\ \text{mol}\ I^-}{\text{L}} \times \frac{1\text{L}}{0.0500\ \text{mol}\ I^-} \times \frac{1\ \text{mL}}{10^{-3}\ \text{L}} = 106.4\ \text{mL}$$

Add 6.4 mL water to the 100.0 mL of the solution.

73. a. Adding H_2SO_4 (aq) will produce a precipitate of barium sulfate, with sodium ion appearing in solution.
b. Water will dissolve sodium carbonate, with magnesium carbonate left as a solid.
c. HCl (aq) will precipitate silver chloride, with potassium ion appearing in solution.
d. Dilute HCl (aq) will dissolve $CuCO_3$, leaving insoluble $PbSO_4$ (s).
e. Dilute HCl (aq) will neutralize the basic magnesium hydroxide. Mg^{2+} (aq) appears in solution and $BaSO_4$ (s) remains undissolved.

75. $2\ HCl + Na_2CO_3 \rightarrow 2\ NaCl + H_2O + CO_2$

$$125\ mL \times \frac{10^{-3}\ L}{mL} \times 1.05\ M\ Na_2CO_3 \times \frac{2\ mol\ NaCl}{mol\ Na_2CO_3} \times \frac{58.44\ g\ NaCl}{mol}$$
$$= 15.3\ g\ NaCl$$

$$75\ mL \times \frac{10^{-3}\ L}{mL} \times 4.5\ M\ HCl \times \frac{2\ mol\ NaCl}{2\ mol\ HCl} \times \frac{58.44\ g\ NaCl}{mol} = 20\ g\ NaCl$$

Na_2CO_3 is the limiting reactant. The mass obtained is 15.3 g NaCl.

77. +7 -2 $S_2O_8^{2-}$

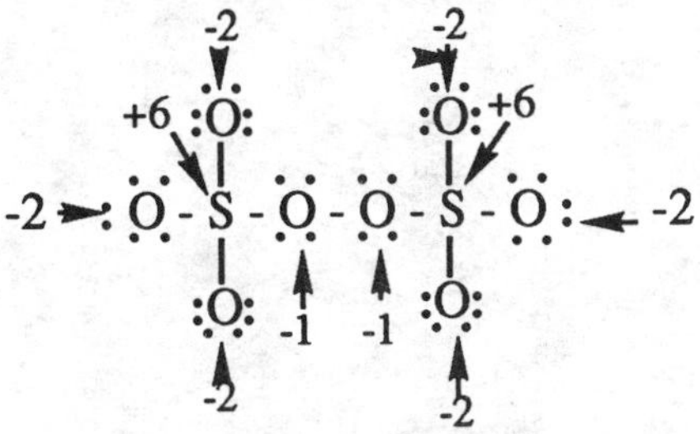

79. $P_4 + 12\ H^+ + 12\ e^- \rightarrow 4\ PH_3$ $P_4 + 16\ H_2O \rightarrow 4\ H_3PO_4 + 20\ H^+ + 20\ e^-$
$5\ P_4 + 3\ P_4 + 48\ H_2O \rightarrow 20\ PH_3 + 12\ H_3PO_4$
$2\ P_4 + 12\ H_2O \rightarrow 5\ PH_3 + 3\ H_3PO_4$

81. a. $I_2 + 6\ H_2O \rightarrow 2\ IO_3^- + 12\ H^+ + 10\ e^-$ $H_5IO_6 + H^+ + 2\ e^- \rightarrow IO_3^- + 3\ H_2O$
$I_2 + 5\ H_5IO_6 \rightarrow 7\ IO_3^- + 9\ H_2O + 7\ H^+$

b. $8\ SCl_2 + 16\ NH_3 + 16\ H^+ + 16\ e^- \rightarrow S_8 + 16\ NH_4Cl$
$4\ SCl_2 + 12\ NH_3 \rightarrow S_4N_4 + 8\ NH_4Cl + 4\ H^+ + 4\ e^-$
$24\ SCl_2 + 64\ NH_3 + \rightarrow S_8 + 4\ S_4N_4 + 48\ NH_4Cl$

c. $XeF_6 + 6\ e^- \rightarrow Xe + 6\ F^-$ $XeF_6 + 6\ H_2O \rightarrow XeO_6^{4-} + 6\ F^- + 12\ H^+ + 2\ e^-$
$2\ H_2O \rightarrow O_2 + 4\ H^+ + 4\ e^-$
$2\ XeF_6 + 8\ H_2O \rightarrow Xe + XeO_6^{4-} + 12\ F^- + O_2 + 16\ H^+$
$2\ XeF_6 + 16\ OH^- \rightarrow Xe + XeO_6^{4-} + 12\ F^- + O_2 + 8\ H_2O$

d. $S_4N_4 + 6\ H_2O + 4\ e^- \rightarrow 2\ S_2O_3^{2-} + 4\ NH_3$
$S_4N_4 + 12\ H_2O \rightarrow 4\ SO_3^{2-} + 4\ NH_3 + 12\ H^+ + 4\ e^-$
$2\ S_4N_4 + 18\ H_2O \rightarrow 2\ S_2O_3^{2-} + 4\ SO_3^{2-} + 8\ NH_3 + 12\ H^+$
$S_4N_4 + 9\ H_2O \rightarrow S_2O_3^{2-} + 2\ SO_3^{2-} + 4\ NH_3 + 6\ H^+$
$S_4N_4 + 3\ H_2O + 6\ OH^- \rightarrow S_2O_3^{2-} + 2\ SO_3^{2-} + 4\ NH_3$

83. $C_{12}H_4Cl_6 + 24\ H_2O \rightarrow 12\ CO_2 + 6\ HCl + 46\ H^+ + 46\ e^-$

$O_2 + 4\ H^+ + 4\ e^- \rightarrow 2\ H_2O$

$2\ C_{12}H_4Cl_6 + 23\ O_2 + 2\ H_2O \rightarrow 24\ CO_2 + 12\ HCl$

Getting rid of toxic wastes instead of leaving them in their molecular form where they might get into the environment is a great advantage. Incineration is one of the very best methods. Sometimes the incinerated material can even produce enough heat to generate electricity.

The major disadvantage of incineration is the possibility that toxic gases might be produced during incineration or that toxic residues may remain.

85. $H_2C_2O_4 + 2\ NaOH \rightarrow Na_2C_2O_4 + 2\ H_2O$

$H_2C_2O_4 \rightarrow 2\ CO_2 + 2\ H^+ + 2\ e^-$ $\quad$ $MnO_4^- + 8\ H^+ + 5\ e^- \rightarrow Mn^{2+} + 4\ H_2O$

$5\ H_2C_2O_4 + 2\ MnO_4^- + 6\ H^+ \rightarrow 10\ CO_2 + 2\ Mn^{2+} + 8\ H_2O$

$$0.03215\ \text{L} \times 0.1050\ \text{M NaOH} \times \frac{\text{mol } H_2C_2O_4}{2\ \text{mol NaOH}} \times \frac{2\ \text{mol } MnO_4^-}{5\ \text{mol } H_2C_2O_4} \times \frac{1}{0.02812\ \text{L}} =$$

$0.02401\ \text{M } KMnO_4$

Chapter 13

Solutions

Exercises

13.1 $\frac{163\text{ g glucose}}{(163\text{ g} + 755\text{ g})\text{ solution}} = 17.8\%$ glucose by mass

13.2 $\frac{40.0\text{ mL ethanol}}{100\text{ mL solution}} \times 200\text{ mL solution} = 80\text{ mL ethanol}$

80 mL ethanol in 120 mL of water.

The assumption is that the volumes will be exactly additive.

13.3 $\frac{0.1\ \mu g}{L} = \frac{0.1\ \mu g}{1000\text{ g}} = \frac{0.1\ \mu g}{1000\text{ g} \times \frac{\mu g}{10^{-6}\text{ g}}} = 0.1\text{ ppb}$

$\frac{0.1\ \mu g}{L} = \frac{0.1\ \mu g}{1000\text{ g}} = \frac{0.1\ \mu g \times 10^3}{1000\text{ g} \times \frac{\mu g}{10^{-6}\text{ g}} \times 10^3} = \frac{100\ \mu g}{10^{12}\ \mu g} = 100\text{ ppt}$

13.4 $\frac{5.00\text{ mL} \times 0.789\text{ g/mL} \times \text{mol}/46.07\text{ g}}{75.0\text{ mL} \times 0.877\text{ g/mL} \times \text{kg}/10^3\text{ g}} = 1.30\text{ m}$

13.5 a. $2.90\text{ mol} \times 32.04\text{ g/mol} = 92.9\text{ g } CH_3OH$

$\frac{92.9\text{ g}}{(\text{kg} \times \frac{1000\text{ g}}{\text{kg}}) + 93\text{ g}} \times 100\% = 8.50\%\ CH_3OH$ by mass

b. $\frac{2.90\text{ mol}}{1093\text{ g soln}} \times \frac{0.984\text{ g}}{\text{mL}} \times \frac{\text{mL}}{10^{-3}\text{ L}} = 2.61\text{ M}$

c. $1000\text{ g } H_2O \times \frac{\text{mol } H_2O}{18.02\text{ g } H_2O} = 55.5\text{ mol}$

$\text{mole }\% = \frac{2.90\text{ mol}}{55.5\text{ mol} + 2.9\text{ mol}} \times 100\% = 4.97\%$

13.6 b. heptane nonpolar dispersion forces only; c. 1-octanol some hydrogen bonding; d. pentanoic acid smaller organic residues, hydrogen bonding; a. acetic acid smallest organic hydrogen bonding

13.7 $\frac{\frac{149\text{ mg } CO_2}{100\text{g } H_2O}}{1\text{ atm } CO_2} \times 0.00036\text{ atm } CO_2 = \frac{5.4 \times 10^{-2}\text{ mg } CO_2}{100\text{ g } H_2O}$

13.8 $1000 \text{ g } H_2O \times \frac{\text{mol}}{18.02 \text{ g}} = 55.5 \text{ mol } H_2O$

$\chi = \frac{55.5 \text{ mol}}{(55.5 + 1.00/\text{mol})} = 0.982$

$P = \chi P^\circ = 0.982 \times 17.5 \text{ mmHg} = 17.2 \text{ mmHg}$

13.9 The greater mole fraction of benzene in the vapor occurs above the solution with the larger mole fraction of benzene. Since toluene has a larger molecular mass, equal masses of toluene and benzene means a greater mole fraction of benzene and a greater mole fraction of benzene in the vapor.

13.10 $\Delta T = 100.35\ ^\circ\text{C} - 100.00\ ^\circ\text{C} = 0.35\ ^\circ\text{C}$

$\Delta T = K_b m$

$m = \frac{\Delta T}{K_b} = \frac{0.35\ ^\circ\text{C}}{0.512\ ^\circ\text{C/m}} = 0.684 \text{ m}$

$\frac{0.684 \text{ mol sucrose}}{\text{kg benzene}} \times \frac{\text{kg}}{1000\text{g}} \times 75.0 \text{ g benzene} \times \frac{342.3 \text{ g}}{\text{mol}} = 17.6 \text{ g sucrose}$

13.11 $\Delta T = 4.25\ ^\circ\text{C} - 5.53\ ^\circ\text{C} = -1.28\ ^\circ\text{C}$

$m = \frac{\Delta T}{K_f} = \frac{-1.28\ ^\circ\text{C}}{-5.12\ ^\circ\text{C/m}} = 0.25 \text{ m}$

$\frac{0.25 \text{ mol cpd}}{\text{kg benz}} \times \frac{\text{kg}}{10^3 \text{ g}} \times 30.00 \text{ g benz} = 0.0075 \text{ mol}$

$\frac{1.065 \text{ g}}{0.0075 \text{ mol}} = 142 \text{ g/mol}$

$50.69 \text{ g C} \times \frac{\text{mol}}{12.01 \text{ g}} = 4.221 \text{ mol C} \times \frac{1}{2.818 \text{ mol}} = 1.50$

$4.23 \text{ g H} + \frac{\text{mol}}{1.008 \text{ g}} = 4.196 \text{ mol H} \times \frac{1}{2.818 \text{ mol}} = 1.49$

$45.08 \text{ g O} \times \frac{\text{mol}}{16.00 \text{ g}} = 2.818 \text{ mol O} \times \frac{1}{2.818 \text{ mol}} = 1.00$

empirical formula: $C_3H_3O_2$ 71 u/formula

$\frac{142 \text{ u}}{\text{molecule}} \times \frac{\text{formula}}{71 \text{ u}} = \frac{2 \text{ formula}}{\text{molecule}}$

molecular formula: $C_6H_6O_4$

13.12 $M = \frac{mRT}{\pi V} = \frac{1.08 \text{ g} \times \frac{62.4 \text{ mmHg L}}{\text{K mol}} \times 298 \text{ K}}{5.85 \text{ mmHg} \times 50.0 \text{ mL} \times \frac{10^{-3} \text{ L}}{\text{mL}}}$

$M = 6.87 \times 10^4 \text{ g/mol}$

Estimation Exercises

13.1 (a). This solution is 15.5% NaCl. The other solutions have less than 15.5 g solute per 100.0 g solution.

13.2 (b). Solution (b) has 25.0% methanol by mass. Solution (a) has 25.0 mL ethanol (25.0 mL x 0.789 g/ml) in 100.0 mL of solution (100.0 mL x 0.968 g/mL); this is only about 20 g solute per 100 g solution -- 20% ethanol by mass.

$$\frac{25.0 \text{ mL ethanol} \times 0.789 \text{ g/mL}}{100 \text{ mL solution} \times 0.968 \text{ g/mL}} \approx 20\% \text{ by mass} \qquad \text{actual } 20.4\ \%$$

13.3 a. $$\frac{1.0 \text{ mol}}{\left(1000 \text{ g} \times \frac{\text{mol}}{18.02}\right) + 1.0 \text{ mol}} = \frac{1.0}{56.5} = 0.018$$

b. $$\frac{10.0 \text{ g} \times \text{mol}/32.04 \text{ g}}{(1000 \text{ g} \times \text{mol}/18.02 \text{ g}) + (10.0\text{g} \times \text{mol}/32.04 \text{ g})} = \frac{0.31}{55.8} = 0.0056$$

c. 0.10

(c) is the greatest mole fraction.

13.4 $\Delta T = -0.12\ ^\circ\text{C} - 0.00\ ^\circ\text{C} = -0.12\ ^\circ\text{C}$

$$m = \frac{\Delta T}{iK_f} = \frac{-0.12\ ^\circ\text{C}}{-3 \times 1.86\ ^\circ\text{C/m}} = 0.022 \text{ m}$$

$MgCl_2 \rightarrow Mg^{2+} + 2\ Cl^-$

Conceptual Exercises

13.1 No. The process will continue until the mole fraction of water in B is equal to the mole fraction in A. The evaporation and condensation will continue but at equal rates in both dishes, and the levels will remain constant.

13.2 The lowest freezing point corresponds to the largest molality of ions. The solutions are dilute enough that molarity is essentially equal to molality.

0.0080 M HCl < 0.0050 m $MgCl_2$ ≈ 0.0030 M $Al_2(SO_4)_3$
0.016 m ions 0.015 m ions 0.015 m ions
lowest f.p.
< 0.010 m $CO(NH_2)_2$
0.010 m particles
highest f.p.

Review Questions

8. $$1\% = \frac{1 \text{ g}}{100\text{g}} \quad \frac{1 \text{ mg}}{\text{dL}} \times \frac{10^{-3} \text{ g}}{\text{mg}} \times \frac{\text{dL}}{10^{-1} \text{ L}} \times \frac{10^{-3} \text{ L}}{\text{mL}} \times \frac{\text{mL}}{1 \text{ g}} = \frac{1 \text{ g}}{10^5 \text{ g}}$$

$$\text{ppt} = \frac{1 \text{ g}}{10^{12} \text{ g}} \quad \text{ppm} = \frac{1 \text{ g}}{10^6 \text{ g}} \quad \text{ppb} = \frac{1 \text{ g}}{10^9 \text{ g}}$$

ppt < ppb < ppm < 1 mg/dL < 1%

12. a. $CHCl_3$ is slightly soluble in water. Although $CHCl_3$ is somewhat polar, the primary intermolecular forces in water are hydrogen bonds, which are unimportant in chloroform ($CHCl_3$).
b. Benzoic acid is slightly soluble in water. There is some hydrogen bonding but the benzene ring is much unlike water.
c. Propylene glycol is highly soluble in water; both have extensive hydrogen bonding.

13. c. Phenol will hydrogen bond to water molecules but also has the benzene ring that makes phenol dissolve in benzene.

24. NaCl produces two ions per formula unit while glucose is a molecular substance. The total particle molarity is the same in the two solutions.

25. a. The cells would shrink. The NaCl (aq) at this concentration is hypertonic.
b. The cells would swell. Osmosis is based on total particle molarity, so 0.92% glucose produces fewer particles than 0.92% NaCl. Glucose has a higher molar mass than NaCl; also, NaCl produces two particles (ions) per formula unit and glucose produces only one.

26. a. 0.10 M $NaHCO_3$ is a higher molarity than 0.05 $NaHCO_3$.
b. 1 M NaCl produces more particles per formula unit than 1 M glucose.
c. 1 M $CaCl_2$ produces more ions per formula unit than 1 M NaCl.
d. 3 M glucose has more particles per liter than 1 M NaCl.

27.

	CH_3OH,	CH_3COOH,	NaCl,	$MgBr_2$,	$Al_2(SO_4)_3$
# particles per formula unit	1	>1	2	3	5
ΔT_f	smallest				largest
T_f	highest				lowest

Problems

29. $2.00 \text{ L} \times \frac{2.00 \text{ L acetic acid}}{100 \text{ L solution}} \times \frac{\text{mL}}{10^{-3} \text{ L}} = 40.0 \text{ mL acetic acid}$

Pipet 40.0 mL of acetic acid into a 2.00 L volumetric flask. Fill with water.

31. a. $\frac{4.12 \text{ g NaOH}}{(100.00 \text{ g} + 4.12 \text{ g}) \text{ solution}} \times 100\% = 3.96\% \text{ by mass}$

b. 5.00 mL x 0.789 g/mL = 3.945 g ethanol

$\frac{3.945 \text{ g ethanol}}{(50.00 \text{ g} + 3.945 \text{ g}) \text{ solution}} \times 100\% = 7.31\% \text{ by mass}$

c. 1.50 mL x 1.324 g/mL = 1.986 g glycerol

22.25 mL x 0.998 g/mL = 22.21 g water

$\frac{1.986 \text{ g glycerol}}{(1.986 \text{ g} + 22.21 \text{ g}) \text{ solution}} \times 100\% = 8.21\% \text{ by mass}$

33. a. $\frac{35.0 \text{ mL } H_2O}{(725 \text{ mL} + 35 \text{ mL}) \text{ solution}} \times 100\% = 4.61\%$ by volume

b. $10.00 \text{ g} \times \frac{\text{mL}}{0.789 \text{ g}} \times \frac{10^{-3} \text{ L}}{\text{mL}} = 0.01267 \text{ L}$

$\frac{0.01267 \text{ L acetone}}{1.55 \text{ L} + 0.013 \text{ L}} \times 100\% = 0.811\%$ by volume

c. $1.05 \text{ g} \times \frac{\text{mL}}{0.810 \text{ g}} = 1.296$ mL 1-butanol

$98.95 \text{ g} \times \frac{\text{mL}}{0.789 \text{ g}} = 125.4$ mL ethanol

$\frac{1.296 \text{ mL 1-butanol}}{(125.4 \text{ mL} + 1.3 \text{ mL}) \text{ solution}} \times 100\% = 1.02\%$ by volume

35. $\frac{0.10 \text{ g glucose}}{100 \text{ g blood}} \times \frac{\text{mg}}{10^{-3} \text{ g}} \times \frac{1.0 \text{ g}}{\text{mL}} \times \frac{\text{mL}}{10^{-3} \text{ L}} \times \frac{10^{-1} \text{ L}}{\text{dL}} = 1 \times 10^2 \text{ mg/dL}$

37. a. $\frac{1 \text{ µg benzene}}{\text{L water}} \times \frac{\text{L}}{10^3 \text{ g}} \times \frac{10^{-6} \text{ g}}{\text{µg}} = \frac{1 \text{ µg}}{10^9 \text{ µg}} = 1$ ppb benzene

b. $\frac{0.0035 \text{ g NaCl}}{100 \text{ g solution}} \times \frac{10^4}{10^4} = \frac{35 \text{ g}}{10^6 \text{ g}} = 35$ ppm NaCl

c. $\frac{2.4 \text{ g } F^-}{10^6 \text{ g solution}} \times \frac{\text{mol}}{19.0 \text{ g}} \times \frac{10^3 \text{ g}}{\text{L}} = 1.3 \times 10^{-4} \text{ M } F^-$

39. $m = \frac{18.0 \text{ g glucose}}{80.0 \text{ g solvent}} \times \frac{10^3 \text{ g}}{\text{kg}} \times \frac{\text{mol}}{180.2 \text{ g}} = 1.25 \text{ m}$

41. $\frac{75 \text{ g } H_3PO_4}{100 \text{ g solution}} \times \frac{1.57 \text{ g}}{\text{mL}} \times \frac{\text{mL}}{10^{-3} \text{ L}} \times \frac{\text{mol}}{97.99 \text{ g}} = 12 \text{ M}$

$\frac{75 \text{ g } H_3PO_4}{25 \text{ g solvent}} \times \frac{10^3 \text{ g}}{\text{kg}} \times \frac{\text{mol}}{97.99 \text{ g}} = 31 \text{ m}$

43. $\frac{3.0 \text{ mol } H_2SO_4}{\text{L solution}} \times \frac{10^{-3} \text{ L}}{\text{mL}} \times \frac{\text{mL}}{1.18 \text{ g solution}} \times \frac{98.09 \text{ g } H_2SO_4}{\text{mol}}$
$\times 100\% = 25\%$ by mass

45. a. $\frac{25.00 \text{ mL methanol x } 0.791 \text{ g/mL}}{25.00 \text{ mL } H_2O \text{ x } 0.998 \text{ g/mL}} \text{ x } \frac{10^3 \text{ g}}{\text{kg}} \text{ x } \frac{\text{mol}}{32.04 \text{ g}} = 24.7 \text{ m}$

b. $\frac{25.00 \text{ mL } H_2O \text{ x } 0.998 \text{ g/mL}}{25.00 \text{ mL methanol x } 0.791 \text{ g/mL}} \text{ x } \frac{10^3 \text{ g}}{\text{kg}} \text{ x } \frac{\text{mol}}{18.02 \text{ g}} = 70.0 \text{ m}$

It should be obvious without the calculation that, because the water has a larger density and a smaller formula weight, that the water solute in methanol solvent will be the larger molality.

The molalities are different because the densities and the molar masses are different. Water in methanol has the larger density in the numerator and also has the smaller molar mass in the denominator.

47. $23.5 \text{ g x } \frac{\text{mol}}{128.16 \text{ g}} = 0.1834 \text{ mol } C_{10}H_8$

$315 \text{ g x } \frac{\text{mol}}{78.11 \text{ g}} = 4.033 \text{ mol } C_6H_6$

$\chi = \frac{0.183 \text{ mol}}{(4.033 + 0.183) \text{ mol}} = 0.0434$

49. $1 \text{ L x } \frac{\text{mL}}{10^{-3} \text{ L}} \text{ x } \frac{0.998 \text{ g}}{\text{mL}} \text{ x } \frac{\text{mol}}{18.02 \text{ g}} = 55.38 \text{ mol } H_2O$

$\chi = \frac{X}{(55.38 + X) \text{ mol}} = \frac{2.50 \text{ mol}}{100 \text{ mol}}$

$100\,X = 138.45 + 2.50\,X$

$97.50\,X = 138.45$

$X = 1.42 \text{ mol}$

mass of sucrose = 1.42 mol x 342.3 g/mol = 486 g

51. $0.10 \text{ m } CO(NH_2)_2 = \frac{0.10 \text{ mol } CO(NH_2)_2}{55.5 \text{ mol } H_2O}$

$\chi = 0.01 = \frac{0.010 \text{ mol } CO(NH_2)_2}{0.990 \text{ mol water}}$

It is obvious without the calculation that because 1000 g H_2O is much larger than 0.99 mol, the $\chi = 0.010$ is greater.

53. a. $\frac{126 \text{ g NaClO}_3}{100\text{g water}}$ $\frac{126 \text{ g NaClO}_3}{226 \text{ g total}} \times 100\% = 55.8\%$ by mass

b. $\frac{33 \text{ g Li}_2\text{SO}_4}{100 \text{ g H}_2\text{O}} \times \frac{1000 \text{ g}}{\text{kg}} \times \frac{\text{mol}}{110 \text{ g}} = 3.0 \text{ m Li}_2\text{SO}_4$

c. $\frac{63.0 \text{ g Pb(NO}_3)_2}{100 \text{ g H}_2\text{O} + 63 \text{ g Pb(NO}_3)_2} = \frac{0.387 \text{ g Pb(NO}_3)_2}{\text{g solution}}$

$\frac{0.387 \text{ g} - 0.325 \text{ g}}{\text{g solution}} = \frac{0.062 \text{ g Pb(NO}_3)_2 \text{ can be added}}{\text{g solution}}$

55. There are a few compounds that are more soluble at colder temperatures.

57. 20° $\frac{2.47 \text{ mg air}}{100\text{g water}}$ 80° $\frac{1.45 \text{ mg air}}{100 \text{ g water}}$

a. $\frac{1.02 \text{ mg air}}{100\text{g H}_2\text{O}} \times 250 \text{ g H}_2\text{O} = 2.55 \text{ mg air}$

b. $$V = \frac{mRT}{PM} = \frac{2.55 \text{ mg} \times \frac{10^{-3} \text{ g}}{\text{mg}} \times \frac{0.08206 \text{ L atm}}{\text{K mol}} \times 273 \text{ K}}{1 \text{ atm} \times 28.96 \text{ g/mol}} = 0.00197 \text{ L} = 1.97 \text{ mL}$$

59. $$P_T = \frac{\frac{2 \text{ g}}{92.13 \text{ g/mol}}}{\frac{2 \text{ g}}{92.13 \text{ g/mol}} + \frac{3\text{g}}{78.11 \text{ g/mol}}} \times 28.4 \text{ mmHg} = 10.3 \text{ mmHg}$$

$$P_B = \frac{\frac{3 \text{ g}}{78.11 \text{ g/mol}}}{\frac{3 \text{ g}}{78.11 \text{ g/mol}} + \frac{2 \text{ g}}{92.13 \text{ g/mol}}} \times 95.1 \text{ mmHg} = 60.8 \text{ mmHg}$$

$$\chi_T = \frac{P_T}{P_{total}} = \frac{10.3 \text{ mmHg}}{(10.3 + 60.8) \text{ mmHg}} = 0.145$$

$$\chi_B = 1.000 - 0.145 = 0.855$$

61. When the solute is non-volatile, the vapor pressure is lower and the boiling point is higher for the solution than for the pure solvent. When the solute is volatile, the vapor pressure of the solution is increased so the boiling point is reduced.

63. a. 0.02 M solution is dilute enough that to 1 significant figure the m = 0.02.

$$\Delta T = K_f \times m = \frac{-1.86\,^\circ\text{C}}{\text{m}} \times 0.02 \text{ m} = -0.037\ ^\circ\text{C}$$

$$\Delta T = t_f - t_i = t_f - 0.00\ ^\circ\text{C} = -0.037\ ^\circ\text{C}$$

$$t_f = -0.04\ ^\circ\text{C}$$

b. $\Delta T = t_f - t_i = 3.85\ ^\circ\text{C} - 5.53\ ^\circ\text{C} = -1.68\ ^\circ\text{C}$

$$\Delta T = K_f \times m \qquad m = \frac{\Delta T}{K_f} = \frac{-1.68\ ^\circ\text{C}}{\frac{-5.12\ ^\circ\text{C}}{\text{m}}}$$

$$m = 0.328$$

c. $\Delta T = K_f \times m = \frac{-1.86\ ^\circ\text{C}}{\text{m}} \times 0.55 \text{ m}$

$$\Delta T = -1.02\ ^\circ\text{C} = t_f - 0.00\ ^\circ\text{C}$$

$$t_f = 1.02\ ^\circ\text{C}$$

$$\Delta T = -1.02\ ^\circ\text{C} - 5.53\ ^\circ\text{C} = -6.55\ ^\circ\text{C}$$

$$m = \frac{\Delta T}{K_f} = \frac{-6.55\ ^\circ\text{C}}{\frac{5.12\ ^\circ\text{C}}{\text{m}}}$$

$$m = 1.28$$

65. 0.08 M is dilute enough to be about 0.08 m to one significant figure.

$$\frac{0.0010 \text{ mol urea}}{0.9990 \text{ mol water}} \times \frac{\text{mol}}{18.02 \text{ g}} \times \frac{10^3 \text{ g}}{\text{kg}} = 0.056 \text{ m}$$

The 0.08 M is more concentrated and thus will have a higher boiling point.

67. $\Delta T = 4.25\ ^\circ\text{C} - 5.53\ ^\circ\text{C} = -1.28\ ^\circ\text{C}$

$$m = \frac{\Delta T}{K_f} = \frac{-1.28\ ^\circ\text{C}}{\frac{-5.12\ ^\circ\text{C}}{\text{m}}} = 0.250 \text{ m}$$

25.00 mL x 0.874 g/mL = 21.85 g benzene

$$21.85 \text{ g} \times \frac{\text{kg}}{10^3 \text{ g}} \times 0.250 \text{ m} = 0.005463 \text{ mol compound}$$

$$\frac{1.45 \text{ g}}{0.005463 \text{ mol}} = 265 \text{ g/mol}$$

69. Because the cucumbers shrivel up, water is leaving the cucumbers, so the salt solution must have a higher osmotic pressure.

71. $$\pi = \frac{mRT}{\mathcal{M}V} = \frac{1.80 \text{ g x } \frac{0.08206 \text{ L atm}}{\text{K mol}} \text{ x } (37 + 273)\text{K}}{46.07 \text{ g/mol x } 100 \text{ mL x } 10^{-3} \text{ L/mL}}$$

$\pi = 9.94$ atm

$$\pi \text{ glucose} = \frac{5.5 \text{ g x } \frac{0.08206 \text{ L atm}}{\text{K mole}} \text{ x } (37 + 273)\text{K}}{180.16 \text{ g/mol x } 100 \text{ mL x } 10^{-3} \text{ L/mL}}$$

$\pi = 7.77$ atm

The CH_3CH_2OH solution is hypertonic.

73. The HCl does not dissociate in benzene. The van't Hoff factor is about 1 for HCl in benzene. $\Delta T_f = -1 \text{ x } 5.12 \text{ x } 0.01 \approx -0.05$ °C. HCl does ionize in water producing a van't Hoff factor of about 2: $\Delta T_f = -2 \text{ x } 1.86 \text{ x } 0.01 \approx -0.04$ °C.

75. $\Delta T = 100.0 \text{ °C} - 99.4 \text{ °C} = 0.6 \text{ °C}$

$$0.6 \text{ °C} = \frac{0.512 \text{ °C}}{\text{m}} \text{ x } m \text{ x } 2 = K_f \text{ x } m \text{ x } i$$

$m = 0.59$ m

$$0.59 \text{ m x } 3.50 \text{ kg x } \frac{58.44 \text{ g}}{\text{mol}} = 1.2 \text{ x } 10^2 \text{ g}$$

77. The lowest freezing point will have the largest ΔT. Instead of calculating ΔT, it is easier to compare i x m, as all of the solutions are in the same solvent, water. The largest product (i x m) produces the lowest freezing point.

b.	1 x 0.15 =	0.15
a.	slightly more than 1 x 0.15 =	slightly more than 0.15
e.	2 x 0.10 =	0.20
c.	slightly more than 2 x 0.10 =	slightly more than 0.20
d.	3 x 0.10 =	0.30

d has the lowest freezing point.

79. $$\frac{1 \text{ cm}}{10 \text{ nm}} \text{ x } \frac{10^{-2} \text{ m}}{\text{cm}} \text{ x } \frac{\text{nm}}{10^{-9} \text{ m}} = 10^6 \text{ cubes on one side}$$

$(10^6)^3 = 10^{18}$ total cubes.

Each cube has (6 x 10 nm x 10 nm =) 600 nm^2 area.

$$600 \text{ nm}^2 \text{ x } 10^{18} \text{ cubes x } \frac{(10^{-9} \text{ m})^2}{\text{nm}^2} = 600 \text{ m}^2$$

81. a. $\frac{11.3 \text{ mL } CH_3OH}{75.0 \text{ mL solution}} \times 100\% = 15.1\%$ by volume

b. $$\frac{11.3 \text{ mL } CH_3OH \times \frac{0.793 \text{ g}}{\text{mL}}}{75.0 \text{ mL solution} \times \frac{0.980 \text{ g}}{\text{mL}}} \times 100\% = 12.2\% \text{ by mass}$$

c. $$\frac{11.3 \text{ mL } CH_3OH \times \frac{0.793 \text{ g}}{\text{mL}}}{75.0 \text{ mL solution}} \times 100\% = 11.9 \text{ mass/volume } \%$$

d. $11.3 \text{ mL} \times \frac{0.793 \text{ g}}{\text{mL}} = 8.961 \text{ g } CH_3OH$

$75.0 \text{ mL} \times \frac{0.980 \text{ g}}{\text{mL}} = 73.5 \text{ g solution}$

$73.5 \text{ g} - 9.0 \text{ g} = 64.5 \text{ g water}$

$8.961 \text{ g } CH_3OH \times \frac{\text{mol}}{32.04 \text{ g}} = 0.2797 \text{ mol}$

$64.5 \text{ g } H_2O \times \frac{\text{mol}}{18.02 \text{ g}} = 3.579 \text{ mol}$

$\text{mole } \% = \frac{0.2797 \text{ mol}}{(3.579 + 0.280) \text{ mol}} \times 100\% = 7.25\%$

83. CO_2 (g) reacts with water to form H_2CO_3 (aq), which is neutralized by the NaOH (aq) to produce Na_2CO_3 (aq). The CO_2 (g) is highly soluble in NaOH (aq) because of this reaction.

85. From Figure 13.8 comes the value of 2.45 mg air/100 g H_2O

$$V = \frac{mRT}{P\mathcal{M}} = \frac{2.45 \text{ mg} \times \frac{10^{-3} \text{ g}}{\text{mg}} \times \frac{0.08206 \text{ L atm}}{\text{K mol}} \times \frac{\text{mL}}{10^{-3} \text{ L}} \times 273.2 \text{ K}}{1 \text{ atm} \times 28.96 \text{ g/mol}} = \frac{1.90 \text{ mL air}}{100\text{g } H_2O}$$

87. Vitamin E is fat soluble, as dispersion forces are the predominant intermolecular forces, both in vitamin E and in fats.
Vitamin B_2 is water soluble, as there are many hydrogen bonding sites.

89. $\chi_{solute} = \frac{n_{solute}}{n_{solute} + n_{solvent}}$. In a dilute solution, $n_{solute} << n_{solvent}$ or $n_{solute} + n_{solvent} \approx n_{solvent}$, and $\chi_{solute} \approx \frac{n_{solute}}{n_{solvent}}$. Because $n_{solvent}$ is proportional to the mass of solvent, χ_{solute} is proportional to molality.

91. Assuming that a liter of solution used 1 kg of solvent, the number of moles of ions can be used to calculate a molality.

$$3.5 \text{ g NaCl} \times \frac{2 \text{ ions}}{\text{formula unit}} \times \frac{\text{mol}}{58.44 \text{ g}} = 0.120 \text{ mol}$$

$$1.5 \text{ g KCl} \times \frac{2 \text{ ions}}{\text{formula unit}} \times \frac{\text{mol}}{74.53} = 0.0402 \text{ mol}$$

$$2.9 \text{ g } Na_3C_6H_5O_7 \times \frac{4 \text{ ions}}{\text{formula unit}} \times \frac{\text{mol}}{258 \text{ g}} = 0.0450 \text{ mol}$$

$$20.0 \text{ g } C_6H_{12}O_6 \times \frac{1 \text{ molecule}}{\text{formula unit}} \times \frac{\text{mol}}{180.2 \text{ g}} = 0.111 \text{ mol}$$

total moles = 0.316 mol
$\Delta T = K_f \times m = -1.86 \times 0.316 \text{ m} = -0.59 \text{ °C}$
This is close to -0.52 °C.

93. $$3.364 \text{ g } CO_2 \times \frac{12.011 \text{ g C}}{44.009 \text{ g } CO_2} = 0.9181 \text{ g C}$$

$$1.377 \text{ g } H_2O \times \frac{2.0158 \text{ g H}}{18.015 \text{ g } H_2O} = 0.1541 \text{ g H}$$

1.684-g sample - 0.918 g C - 0.154 g H = 0.612 g O

$$0.918 \text{ g C} \times \frac{\text{mol}}{12.01 \text{ g C}} = 0.07644 \text{ mol C} \times \frac{1}{0.03825 \text{ mol O}} = 2.00$$

$$0.154 \text{ g H} \times \frac{\text{mol}}{1.008 \text{ g H}} = 0.1528 \text{ mol H} \times \frac{1}{0.03825 \text{ mol O}} = 3.99$$

$$0.612 \text{ g O} \times \frac{\text{mol}}{16.00 \text{ g O}} = 0.03825 \text{ mol H} \times \frac{1}{0.03825 \text{ mol O}} = 1.00$$

C_2H_4O formula weight = 44 u

$\Delta T = K_f \times m$

$$m = \frac{\Delta T}{K_f} = \frac{-0.244 \text{ °C}}{-1.86 \text{ °C/m}} = 0.1312 \text{ m}$$

$$0.1312 \text{ m} \times \frac{\text{kg}}{10^3 \text{ g}} \times 34.89 \text{ g} = 0.004577 \text{ mol}$$

$$\mathcal{M} = \frac{0.605 \text{ g}}{0.004577 \text{ mol}} = 132 \text{ g/mol; molecular weight} = 132 \text{ u}$$

$$\frac{132 \text{ u}}{\text{molecule}} \times \frac{\text{formula unit}}{44 \text{ u}} = \frac{3 \text{ formula unit}}{\text{molecule}}$$

molecular formula = 3 x empirical formula = $C_6H_{12}O_3$

Chapter 14

Chemicals In Earth's Atmosphere

Exercises

14.1 $\frac{38.5\%}{100\%} \times 17.5 \text{ mmHg} = 6.74 \text{ mmHg}$

14.2 a. $NH_3\ (aq) + HNO_3\ (aq) \rightarrow NH_4NO_3\ (aq)$
b. $2\ NH_3\ (aq) + H_2SO_4\ (aq) \rightarrow (NH_4)_2SO_4\ (aq)$

Estimation Exercises

14.1 $(NH_4)_2SO_4$ $\qquad (NH_4)_2HPO_4$

%N $\quad \frac{2N}{2N + 8H + S + 4O} \qquad \frac{2N}{2N + 8H + H + P + 4O}$

If we eliminate like elements, the difference is S (32.06) versus H + P (1.008 + 30.97). DAP will have a slightly greater %N because S is slightly heavier than H + P.

14.2 $V_{particle} = \frac{4}{3}\pi r^3 = \frac{4}{3}\pi(0.5\ \mu m)^3$
$= 0.52\ \mu m^3$

$$\frac{100\ \mu g}{m^3 \text{ air}} \times \left(\frac{10^{-2}\ m}{cm}\right)^3 \times \frac{10^{-6}\ g}{\mu g} \times \frac{cm^3}{1\ g} \times \frac{\text{particle}}{0.52\ \mu m^3} \times \left(\frac{10^{-2}\ m}{cm}\right)^3 \times \left(\frac{\mu m}{10^{-6}\ m}\right)^3 =$$

2×10^2 particles/cm^3 air

Conceptual Exercise

14.1 $1\ L \times \frac{mL}{10^{-3}\ L} \times \frac{0.80\ g}{mL} \times \frac{mol}{114.22\ g} \times \frac{8\ mol\ CO}{mol\ C_8H_{18}} \approx 56\ mol\ CO$

C_8H_{18} is less dense than water; that is $d < 1.00$ g/mL. Let's assume about 0.80 g/mL.

$$95\ m \times 38\ m \times 16\ m \times \left(\frac{cm}{10^{-2}\ m}\right)^3 \times \frac{mL}{cm^3} \times \frac{10^{-3}\ L}{mL} \times \frac{1\ mol\ air}{25\ L\ air} \approx 2.3 \times 10^6\ mol$$

Assume a molar volume of air of about 25 L/mol at the prevailing T and P.

$$\frac{56\ mol\ CO}{2.3 \times 10^6\ mol\ air} \approx 24\ ppm$$

The limit of 35 ppm would not be exceeded.

Review Questions

10. a. anhydrous ammonia
    b. carbon monoxide, nitrogen monoxide, and carbon dioxide
    c. methane, ozone, nitrous oxide, CFCs
    d. radon
    e. nitric acid, and sulfuric acid
    f. nitrogen, helium, argon
    g. helium and oxygen

11. a. magnesium nitride
    b. N_2O_4
    c. potassium peroxide
    d. KO_2
    e. 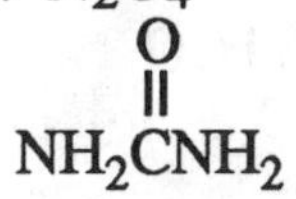

22. a. Nitrogen monoxide reacts with oxygen to produce NO_2, and it also participates in reactions with hydrocarbons that lead to ozone, PAN, and other smog components.
    b. Carbon monoxide is not important in formation of photochemical smog.
    c. Hydrocarbon vapors react with oxygen atoms to make free radicals, ozone and PAN.
    d. Sulfur oxide is not important in photochemical smog. It is important in industrial smog.

Problems

29. argon $\frac{0.934\text{ L}}{100\text{ L}} \times \frac{10^4}{10^4} = \frac{9.34 \times 10^3\text{ L}}{10^6} = 9.34 \times 10^3\text{ ppm}$

neon $\frac{0.001818}{100\text{ L}} \times \frac{10^4}{10^4} = \frac{18.18\text{ L}}{10^6} = 18.18\text{ ppm}$

helium $\frac{0.000524}{100\text{ L}} \times \frac{10^4}{10^4} = \frac{5.24\text{ L}}{10^6} = 5.24\text{ ppm}$

krypton $\frac{0.000114}{100\text{ L}} \times \frac{10^4}{10^4} = \frac{1.14\text{ L}}{10^6} = 1.14\text{ ppm}$

xenon $\frac{0.000009}{100\text{ L}} \times \frac{10^4}{10^4} = \frac{0.09\text{ L}}{10^6} = 0.09\text{ ppm}$

31. $\text{R.H.} = \frac{10.5\text{ mmHg}}{23.8\text{ mmHg}} \times 100\% = 44.1\%$

33. It has the greatest relative humidity at 25 °C. Raising the temperature raises the vapor pressure of water and thus lowers the relative humidity.

35. Steam is visible when the warm moist air is mixing with cooler air. The absolute humidity of air above a kettle of boiling water is high--the air is nearly saturated in water vapor. The temperature must be reduced only slightly to reach the dew point. The absolute humidity in expired air is much less, and the temperature has to be lowered much further to reach the dew point. Only on a cold day will the partial pressure of water vapor in your breath be high enough for condensation to occur.

37. a. N_2O $\frac{28}{28+16} = \frac{14}{14+8}$ b. NH_3 $\frac{14}{14+3}$

c. NO $\frac{14}{14+16}$ d. NH_4Cl $\frac{14}{14+4+35}$

$NH_4Cl < NO < N_2O < NH_3$

39. a. $2\ C_8H_{18} + 25\ O_2 \rightarrow 16\ CO_2 + 18\ H_2O$
b. $2\ CH_4 + 3\ O_2 \rightarrow 2\ CO + 4\ H_2O$
c. $2\ CO + O_2 \xrightarrow{Pt} 2\ CO_2$

41. Haber-Bosch $N_2\ (g) + 3H_2\ (g) \xrightarrow{catalyst} 2NH_3\ (g)$
followed by the Ostwald process
Ostwald $4\ NH_3\ (g) + 5\ O_2\ (g) \xrightarrow{Pt/Rh} 4\ NO\ (g) + 6\ H_2O\ (g)$
$2\ NO\ (g) + O_2\ (g) \rightarrow 2\ NO_2\ (g)$
$3\ NO_2\ (g) + H_2O\ (l) \rightarrow 2\ HNO_3\ (aq) + NO\ (g)$
$NH_3\ (aq) + HNO_3\ (aq) \rightarrow NH_4NO_3\ (aq)$

43. $4000\ g\ N \times \frac{132.14\ g\ (NH_4)_2SO_4}{2 \times 14.01\ g\ N} \times \frac{10^{-3}\ kg}{1\ g} = 18.9\ kg$

45. a. $4\ Al\ (s) + 3\ O_2\ (g) \rightarrow 2\ Al_2O_3\ (s)$
b. $2\ KClO_3\ (s) \xrightarrow{\Delta} 2\ KCl\ (s) + 3\ O_2\ (g)$
c. $2\ H_2O + 2\ Na_2O_2\ (s) \rightarrow 4\ NaOH\ (aq) + O_2\ (g)$

47. $4\ KO_2\ (s) + 2\ CO_2\ (g) \rightarrow 2\ K_2CO_3\ (s) + 3\ O_2\ (g)$
The pressure will be increased because two moles of gas make three moles of gas.

49. Combustion of a hydrocarbon is an oxidation-reduction reaction in which the oxidation state of carbon atoms can increase to either +2 (CO) or +4 (CO_2). The reaction of an acid with a metal carbonate is an acid-base reaction. The oxidation state of C is +4 in the metal carbonate and remains the same in CO_2. A reduction would be required to produce CO, but there is no accompanying oxidation.

51. $2\ C_6H_{14} + 19\ O_2 \rightarrow 12\ CO_2 + 14\ H_2O$
It is impossible because no one can know the ratio of CO to CO_2 produced at any one time. The ratio can even change during the reaction.

53. a. $CH_4 + 2\,O_2 \rightarrow CO_2 + 2\,H_2O$

$$19.8\text{ t} \times \frac{10^3\text{ kg}}{1\text{ t}} \times \frac{10^3\text{ g}}{1\text{ kg}} \times \frac{\text{mol } CO_2}{44.01\text{ g } CO_2} \times \frac{\text{mol } CH_4}{\text{mol } CO_2} \times \frac{16.04\text{ g } CH_4}{\text{mol } CH_4} \times \frac{10^{-3}\text{ kg}}{1\text{ g}} \times \frac{10^{-3}\text{ t}}{1\text{ kg}} = 7.22\text{ metric tons } CH_4$$

b. $2\,C_8H_{18} + 25\,O_2 \rightarrow 16\,CO_2 + 18\,H_2O$

$$19.8\text{ t} \times \frac{10^3\text{ kg}}{1\text{ t}} \times \frac{10^3\text{ g}}{1\text{ kg}} \times \frac{\text{mol } CO_2}{44.01\text{ g } CO_2} \times \frac{2\text{ mol } C_8H_{18}}{16\text{ mol } CO_2} \times \frac{114.22\text{ g } C_8H_{18}}{\text{mol } C_8H_{18}} \times \frac{10^{-3}\text{ kg}}{1\text{ g}} \times \frac{10^{-3}\text{ t}}{1\text{ kg}} = 6.42\text{ metric tons } C_8H_{18}$$

c. $C + O_2 \rightarrow CO_2$

$$19.8\text{ t} \times \frac{10^3\text{ kg}}{1\text{ t}} \times \frac{10^3\text{ g}}{1\text{ kg}} \times \frac{\text{mol } CO_2}{44.01\text{ g } CO_2} \times \frac{\text{mol C}}{\text{mol } CO_2} \times \frac{12.01\text{ g C}}{\text{mol C}} \times \frac{100\text{ g coal}}{94.1\text{ g C}} \times \frac{10^{-3}\text{ kg}}{1\text{ g}} \times \frac{10^{-3}\text{ t}}{1\text{ kg}} = 5.74\text{ metric tons coal}$$

55. a. $S\,(s) + O_2\,(g) \rightarrow SO_2\,(g)$
b. $2\,ZnS\,(s) + 3\,O_2\,(g) \rightarrow 2\,ZnO\,(s) + 2\,SO_2\,(g)$
c. $2\,SO_2\,(g) + O_2\,(g) \rightarrow 2\,SO_3\,(g)$
d. $SO_3\,(g) + H_2O\,(l) \rightarrow H_2SO_4\,(aq)$
e. $H_2SO_4\,(aq) + 2\,NH_3\,(aq) \rightarrow (NH_4)_2SO_4\,(aq)$

57. $$1\text{ t coal} \times \frac{5\text{ g S}}{100\text{ g coal}} \times \frac{\text{mol S}}{32.07\text{ g S}} \times \frac{\text{mol } SO_2}{\text{mol S}} \times \frac{64.07\text{ g } SO_2}{\text{mol } SO_2} = 0.0999\text{ t } SO_2.$$

Even assuming a high value of 5% S in coal or natural gas, the ZnS smelter still produces more SO_2. In smelting of ZnS and PbS, all the sulfur is converted to SO_2. Likewise, the sulfur content of coal and natural gas will appear as SO_2. The material that will produce the greatest quantity of SO_2 per ton is the one that has the highest % S. The typical natural gas is very low in sulfur content. Coal may have about 5% or so. The % S in ZnS is greater than in PbS because the atomic weight of Zn is much less than that of Pb. The smelting of ZnS produces the greatest quantity of SO_2.

$2\,ZnS + 3\,O_2 \rightarrow 2\,ZnO + 2\,SO_2$

$$1\text{ t ZnS} \times \frac{\text{mol ZnS}}{97.46\text{ g ZnS}} \times \frac{2\text{ mol } SO_2}{2\text{ mol ZnS}} \times \frac{64.07\text{ g } SO_2}{\text{mol } SO_2} = 0.657\text{ t } SO_2$$

$2\,PbS + 3\,O_2 \rightarrow 2\,PbO + 2\,SO_2$

$$1\text{ t PbS} \times \frac{\text{mol PbS}}{239.3\text{ g PbS}} \times \frac{2\text{ mol } SO_2}{2\text{ mol PbS}} \times \frac{64.06\text{ g } SO_2}{\text{mol } SO_2} = 0.268\text{ t } SO_2$$

59. 0.3% volume is the same as 0.3% mole. The mole percent He in air is 0.000524. So the He in natural gas is about 600 times more abundant than in air, that is, $0.3/0.000524 \approx 600$.

61. Helium is formed from an alpha particle. There are many reactions involving an alpha particle. Argon is formed only when the potassium-40 isotope decays.

63. $\text{R.H.} = \frac{5.67 \text{ mmHg}}{17.5 \text{ mmHg}} \times 100\% = 32.4\%$

It does take some moisture out of the air, but only if the initial relative humidity is greater than 32.4%.

65. $$\frac{\chi_{N_2}}{\chi_{N_2} + \chi_{Ar}} \quad \frac{78.084 \text{ mol } N_2}{78.084 \text{ mol } N_2 + 0.934 \text{ mol Ar}} = \frac{78.084}{79.018} \times 100\% = 98.82\% \; N_2$$

$$\frac{98.82\% \; N_2}{100\%} \times 28.02 + \frac{1.18 \; \% \text{ Ar}}{100\%} \times 39.95 = 28.16 \text{ apparent molar mass}$$

That is, when all other gases are removed from air, the remaining N_2-Ar mixture is 98.82 mole % N_2 and 1.18 mole % Ar.

$$d_{N2\text{-}Ar} = \frac{28.16 \text{ g}}{22.414 \text{ L}} = 1.256 \text{ g/L (at STP)}$$

$$d_{N2} = \frac{28.02 \text{ g}}{22.414 \text{ L}} = 1.250 \text{ g/L (at STP)}$$

$$\% \text{ difference} = \frac{1.256\text{-}1.250}{1.250} \times 100 = 0.48 \; \%$$

67. $$2500 \text{ ton} \times \frac{2000 \text{ lb}}{\text{ton}} \times \frac{454 \text{ g}}{\text{lb}} \times \frac{0.65 \text{ g S}}{100\text{g coal}} \times \frac{\text{mol S}}{32.07 \text{ g S}} \times \frac{\text{mol } SO_2}{\text{mol S}} \times \frac{64.07 \text{ g } SO_2}{\text{mol } SO_2}$$

$$= 2.95 \times 10^7 \text{ g } SO_2$$

$V = 45 \text{ km} \times 60 \text{ km} \times 0.40 \text{ km} = 1080 \text{ km}^3$

$$\frac{2.94 \times 10^7 \text{ g } SO_2}{1080 \text{ km}^3} \times \left(\frac{\text{km}}{10^3 \text{ m}}\right)^3 \times \frac{\mu\text{g}}{10^{-6} \text{ g}} = 27.3 \frac{\mu\text{g } SO_2}{\text{m}^3 \text{ air}}$$

No, it will not exceed the standard.

69. $$5.2 \times 10^{15} \text{ metric ton} \times \frac{1000 \text{ kg}}{\text{metric ton}} \times \frac{10^3 \text{ g}}{\text{kg}} \times \frac{\text{mol air}}{28.97 \text{ g}} \times \frac{359 \text{ mole } CO_2}{10^6 \text{ mol air}}$$

$$\times \frac{44.01 \text{ g } CO_2}{\text{mol } CO_2} = 2.84 \times 10^{18} \text{ g } CO_2$$

$$\frac{4.6 \times 10^{16} \text{ g } CO_2}{2.84 \times 10^{18} \text{ g } CO_2} \times 100\% = 1.6 \; \% \text{ increase}$$

molar mass of air = 28.0134 x 0.78084 + 32.000 x 0.20946 + 39.948 x 0.00934 + 44.009 x 0.000356 = 28.966 g/mol

Chapter 15

Chemical Kinetics

Exercises

15.1 a. $\frac{-\Delta[B]}{\Delta t} = \frac{1}{2} \times \frac{-\Delta[A]}{\Delta t} = \frac{1}{2} \times 2.10 \times 10^{-5}$ M/s = 1.05×10^{-5} M/s

b. $\frac{1\Delta[C]}{3\ \Delta t} = \frac{-1\Delta[A]}{2\ \Delta t}$

$\frac{\Delta[C]}{\Delta t} = \frac{3}{2} \times 2.10 \times 10^{-5}$ M/S = 3.15×10^{-5} M/s

15.2 a. From tangent line

rate = slope = $\frac{-(0 - 0.63)\text{ M}}{(570 - 0)\text{ s}} = 1.11 \times 10^{-3}$ M/s

b. $\Delta[H_2O_2] = -10\text{ s} \times 1.11 \times 10^{-3}\text{ M/s} = -1.1 \times 10^{-2}$ M

$[H_2O_2]_{310} = [H_2O_2]_{300} + \Delta[H_2O_2]$

$= 0.298\text{ M} + (-1.1 \times 10^{-2}\text{ M}) = 0.287$ M

15.3 rate = $k\,[NO]^2[O_2]$

rate = $\frac{7.05 \times 10^3}{M^2\ s} \times (0.200\text{ M})^2 \times 0.400$ M

rate = 113 M/s

15.4 $\ln\frac{[H_2O_2]_t}{[H_2O_2]_o} = -kt$

$\ln\frac{0.500}{0.882} = \frac{-3.66 \times 10^{-3}}{s}\,t$

$-0.568 = \frac{-3.66 \times 10^{-3}}{s}\,t$

$t = 155$ s

15.5 $\frac{M_t}{M_o} = \frac{\frac{n_t}{V}}{\frac{n_o}{V}} = \frac{n_t}{n_o} = \frac{\frac{m_t}{\mathcal{M}}}{\frac{m_o}{\mathcal{M}}} = \frac{m_t}{m_o}$

$\ln \frac{m_t}{m_o} = -kt$

$\ln \frac{m_t}{45.0 \text{ g}} = \frac{-5.83 \times 10^{-3}}{\text{s}} \times 5.00 \text{ min} \times 60 \text{ s/min} = -1.749$

$\frac{m_t}{45.0 \text{ g}} = e^{-1.749} = 0.174$

$m_t = 7.83 \text{ g}$

15.6 Because $\frac{1}{32}$ is $\left(\frac{1}{2}\right)5$, the time required is 5 half-lives, that is, 5 x $t_{1/2}$.

5 x 118 s = 590 s

or

$\ln \frac{\frac{1}{32}}{1} = \frac{-5.83 \times 10^{-3}}{\text{s}} t$

$-3.47 = \frac{-5.83 \times 10^{-3}}{s} t$

$t = 594 \text{ s}$

15.7 $\ln \frac{A_t}{A_o} = -\lambda t$

$\lambda = \frac{0.693}{t_{1/2}} = \frac{0.693}{2.411 \times 10^4 \text{ y}}$

$\lambda = \frac{2.874 \times 10^{-5}}{\text{y}}$

$\ln \frac{1}{100} = \frac{-2.874 \times 10^{-5}}{\text{y}} t$

$-4.61 = \frac{-2.874 \times 10^{-5}}{\text{y}} t$

$t = 1.60 \times 10^5 \text{ y}$

15.8 The activity falls to one-half its initial value in one half-life period -- $t_{1/2}$ = 12.26 y. The brandy is 12.26 y old, not 25 years. The claimed age is not authentic.

15.9 The reaction begins at $\frac{1}{[A]_o} = 1.065\ M^{-1}$

$[A]_o = 1/1.065\ M^{-1} = 0.9390\ M$

$t_{1/2} = \frac{1}{k} \times \frac{1}{[A]_o} = \frac{1}{2.3 \times 10^{-4}\ M^{-1}.\ min^{-1}} \times 1.065\ M^{-1} = 4.63 \times 10^3\ min;$

$[A]_{1/2} = \frac{1}{2} \times 0.9390\ M = 0.4695\ M$

$t_{1/4} = \frac{1}{2.3 \times 10^{-4}\ M^{-1}.\ min^{-1}} \times \frac{1}{0.4695\ M} = 9.26 \times 10^3/min$

$t_{1/4} = 2 \times 4.63 \times 10^3\ min = 2 \times t_{1/2}$

15.10 $\ln \frac{1.0 \times 10^{-5}}{2.5 \times 10^{-3}} = \frac{1.0 \times 10^2\ kJ/mol \times 1000\ J/kJ}{\frac{8.3145\ J}{mol\ K}} \times \left(\frac{1}{332\ K} - \frac{1}{T_2}\right)$

$-5.52 = 1.20 \times 10^4\ K^{-1} \times \left(\frac{1}{332\ K} - \frac{1}{T_2}\right)$

$-4.60 \times 10^{-4}\ K^{-1} = \frac{1}{332\ K} - \frac{1}{T_2}$

$\frac{1}{T_2} = \frac{1}{332\ K} + 4.60 \times 10^{-4}\ K^{-1} = 3.01 \times 10^{-3}\ K^{-1} + 4.60 \times 10^{-4}\ K^{-1} =$
$3.47 \times 10^{-3}\ K^{-1}$

$T_2 = 1/3.47 \times 10^{-3}\ K^{-1} = 288\ K\ (15\ °C)$

15.11 a. Net reaction $2\ NO + Cl_2 \rightarrow 2\ NOCl$

b. $NOCl_2$ is an intermediate.

c. The first step is rate determining. Its reactants are in the rate law.

Estimation Exercises

15.1 a. rate = - slope = $\frac{-(0.33 - 0.882)\ M}{(420 - 0)\ s}$

rate = 1.3×10^{-3} M/s

Alternate: use initial and final values in Table 15.1

$\frac{-(0.094 - 0.0882)\ M}{(600 - 0)\ s} = 1.3 \times 10^{-3}\ M\ s^{-1}$

b. at about 260 s
Only one tangent line is possible.

15.2 10% is left. That is between 3 $\left(\left(\frac{1}{2}\right)^3 = \frac{1}{8}\right)$ and 4 $\left(\left(\frac{1}{2}\right)^4 = \frac{1}{16}\right)$ half-lives or between 354 s and 472 s. Answer (c)

$\ln \frac{0.1}{1} = -5.83 \times 10^{-3}\ t$

$t = 395$ sec

15.3 $\frac{168 \text{ hr}}{14.569 \text{ hs}} = 11.5$ half-lives

$\left(\frac{1}{2}\right)^{11} = 4.88 \times 10^{-4}$

$\left(\frac{1}{2}\right)^{12} = 2.44 \times 10^{-4}$

$4.88 \times 10^{-4} \times 3.29 \times 10^{14}$ atoms $= 1.61 \times 10^{11}$ atoms

$2.44 \times 10^{-4} \times 3.29 \times 10^{14}$ atoms $= 8.03 \times 10^{10}$ atoms

The answer 1.19×10^{11} does fall between the numbers of 1.61×10^{11} and 8.03×10^{10}.

Conceptual Exercise

t	0	200	400	800
[A]	1.00	0.82	0.71	0.54
$\frac{1}{[A]}$	1	1.22	1.41	1.85

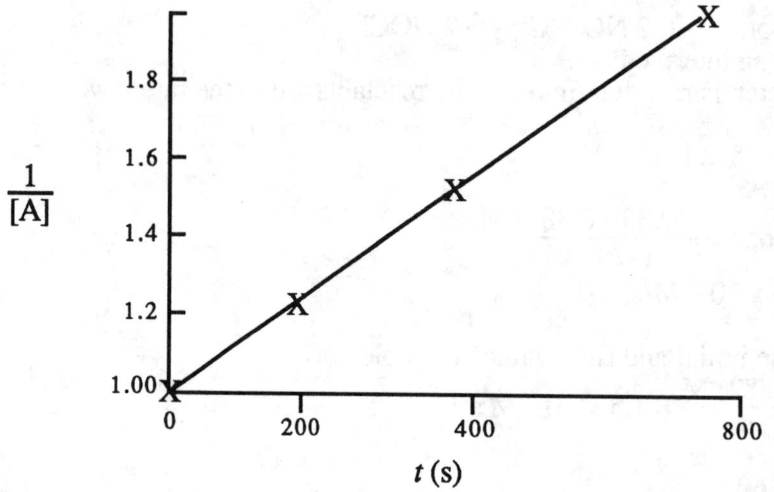

Review Questions

3. Only one O_2 molecule is formed for every two H_2O_2 molecules that react. The rate of formation of O_2 is 1/2 of the rate of disappearance of H_2O_2.

8. It must be a second-order reaction, which has the rate law: rate of reaction = $k[A]^2$. When $[A] = \frac{1}{2}[A_o]$, rate of reaction $= k(\frac{1}{2}[A_o])^2 = \frac{1}{4}k\cdot[A_o]^2$.

12. The one with a short half-life has more activity because half of its atoms must disintegrate in a shorter time than the isotope of longer half-life.

17. a. Z b. X c. the transition state (activated complex) d. activation energy for the forward reaction e. activation energy for the reverse reaction.

30. a. When the substrate concentration is low and is doubled, the rate doubles.
b. When the substrate concentration is high compared to the amount of enzyme, the rate stays constant, even if the substrate concentration is doubled, because all of the enzyme is in use.

Problems

33. $[A] = 0.2546\text{ M} - 4.0 \times 10^{-5}\text{ M/s} \times 35\text{ s} = 0.2532\text{ M}$

35. a. 3.1×10^{-4} M/s $\quad -\frac{1}{2}\frac{d[B]}{dt} = \frac{d[A]}{dt}$

b. 9.3×10^{-4} M/s $\quad -\frac{1}{2}\frac{d[B]}{dt} = -\frac{1}{3}\frac{d[D]}{dt}$

37. rate = k This is a true statement for zero-order reactions.

39. rate = $k[A]^x[B]^y$

Expt. $\frac{4}{1}$ $\quad \frac{(0.76)^x(0.24)^y}{(0.38)^x(0.24)^y} = \frac{2.8 \times 10^{-3}}{2.8 \times 10^{-3}}$

$2^x = 1 \qquad x = 0$

Expt. $\frac{1}{2}$ $\quad \frac{(0.24)^y}{(0.12)^y} = \frac{2.8 \times 10^{-3}}{7.0 \times 10^{-4}}$

$2^y = 4 \qquad y = 2$

a. zero-order in A, second-order in B
b. second-order overall
c. $2.8 \times 10^{-3}\text{ M/s} = k(0.24\text{ M})^2$

$$k = \frac{0.049}{\text{M s}} = 0.049\text{ M}^{-1}\cdot\text{s}^{-1}$$

41. $[H_2O_2] = \frac{3\text{ g } H_2O_2}{100\text{ g solution}} \times \frac{1.00\text{ g}}{\text{mL}} \times \frac{\text{mL}}{10^{-3}\text{ L}} \times \frac{\text{mol}}{34.02\text{ g}} = 0.9\text{ M}$

$$\text{rate} = \frac{3.66 \times 10^{-3}}{\text{s}} \times 0.9\text{ M} = 3 \times 10^{-3}\text{ M/s}$$

43. $[A] = \frac{\text{rate}}{k} = \frac{0.0150\text{ M/min}}{0.0462\text{/min}} = 0.325\text{ M}$

45. a. $\ln \frac{0.264}{0.280} = -k \times 10.0 \text{ min}$ $\quad \ln \frac{0.197}{0.280} = k \times 60.0 \text{ min}$

$k = 5.88 \times 10^{-3}/\text{min}^{-1}$ $\quad k = 5.86 \times 10^{-3} \text{ min}^{-1}$

average value: $5.87 \times 10^{-3} \text{ min}^{-1}$

b. $t_{1/2} = \frac{0.693}{5.87 \times 10^{-3}/\text{min}} = 118 \text{ min}$

c. $\ln \frac{0.0350}{0.280} = -5.87 \times 10^{-3}/\text{min} \times t$

$t = 354 \text{ min}$

47. a. $k = \frac{0.693}{27 \text{ min}} = 2.57 \times 10^{-2}/\text{min}$

$\ln \frac{P}{626 \text{ mmHg}} = -2.57 \times 10^{-2}/\text{min} \times 1.00 \text{ hr} \times \frac{60 \text{ min}}{\text{hr}} = -1.54$

$\frac{P}{626 \text{ mmHg}} = e^{-1.54} = 0.214$

$P = 134 \text{ mmHg}$

b. $P_{CH_4} = P_{H_2} = P_{CO} = 626 \text{ mm Hg} - 134 \text{ mmHg} = 492 \text{ mmHg}$

$P_{total} = P_{CH_4} + P_{H_2} + P_{CO} + P_{(CH_3)_2O}$

$P_{total} = 492 \text{ mmHg} + 492 \text{ mmHg} + 492 \text{ mmHg} + 134 \text{ mmHg} = 1610 \text{ mmHg}.$

49. It is not possible chemically because the energetics of the nucleus are not affected by chemical reactions.

51. For less than a few hundred years old, there is not enough difference between the present activity and the activity of the object. After 50,000 years, there is so little activity left that it is almost impossible to get an accurate measurement.

53. $1.00 \ \mu g \times \frac{10^{-6} \text{ g}}{\mu g} \times \frac{\text{mol}}{239 \text{ g}} \times \frac{6.022 \times 10^{23} \text{ atoms}}{\text{mol}} = 2.52 \times 10^{15} \text{ atoms}$

$\lambda = \frac{2.27 \times 10^{3} \text{ atoms/s}}{2.52 \times 10^{15} \text{ atoms}} = 9.01 \times 10^{-13}/\text{sec}$

$t_{1/2} = \frac{0.693}{9.01 \times 10^{-13} \text{ sec}} = 7.69 \times 10^{11} \text{ sec}$

$= 1.28 \times 10^{10} \text{ min} = 2.14 \times 10^{8} \text{ hr} = 8.90 \times 10^{6} \text{ day}$

$= 2.44 \times 10^{4} \text{ y}.$

55. $1 \rightarrow \frac{1}{2} \rightarrow \frac{1}{4} \rightarrow \frac{1}{8} \rightarrow \frac{1}{16}$

8 + 8 + 8 + 8

More than 24 d and less than 32.

57. $[NH_3]_{335} = 0.0452\ M - 3.40 \times 10^{-6}\ M/s \times 335\ sec = 0.0441\ M$

59. $\frac{1}{[A]_t} = kt + \frac{1}{[A]_o}$

$$\frac{1}{[A]_t} = \frac{2.3 \times 10^{-4}}{M\ min} \times 2.00\ h \times \frac{60\ min}{hr} + \frac{1}{0.56\ M}$$

$$\frac{1}{[A]_t} = \frac{2.76 \times 10^{-2}}{M} + \frac{1.79}{M}$$

$[A]_t = 0.55\ M$

61. For zero-order the half-life gets longer as the initial concentration increases, because the rate is constant and there are more molecules to react.
For second-order, the half-life is related to the inverse of the concentration. An increase in concentration decreases the half-life.

63. An increase in temperature causes, not only the average kinetic energies of the molecules to increase, causing more collisions, but it also causes many more of the collisions to have enough energy to overcome the activation energy.

65.

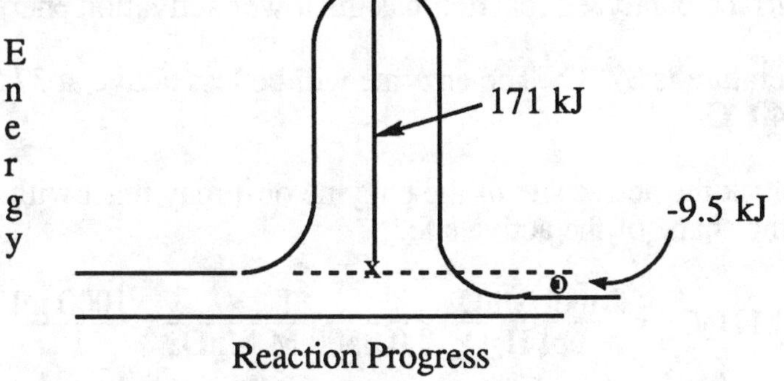

67. $k = \frac{0.693}{t_{1/2}}$ $\quad k_1 = \frac{0.693}{320\ min} = 2.17 \times 10^{-3} min^{-1}$

$k_2 = \frac{0.693}{100\ min} = 6.93 \times 10^{-3}\ mm^{-1}$

$T_1 = 135 + 273 = 408\ K \qquad T_2 = 145 + 273 = 418\ K$

$$\ln\frac{6.93 \times 10^{-3}}{2.17 \times 10^{-3}} = \frac{E_a \times 1000\ J/kJ}{8.3145\ J/mol\ K} \times \left(\frac{1}{408\ K} - \frac{1}{418\ K}\right)$$

$1.16 = E_a \times 120.3\ mol/kJ \times 5.86 \times 10^{-5}$
$E_a = 165\ kJ/mol$

69. $\ln \frac{0.0120}{0.0100} = \frac{218 \text{ kJ/mol} \times 1000 \text{ J/kJ}}{8.3145 \text{ J/mol K}} \times \left(\frac{1}{T_1} - \frac{1}{652 \text{ K}}\right)$

$6.954 \times 10^{-6} = \left(\frac{1}{T_1} - \frac{1}{652 \text{ K}}\right)$

$T_1 = 649$ K

71. The rate equations for an elementary step and for an overall reaction are not likely to be the same unless (a) the reaction occurs in a single step, or (b) the mechanism consists of a slow first step followed by a fast second step. If the rate-determining step is not the first step, the rate equation for the overall reaction is not likely to be the same as that of an elementary step. Also, the rate equations for some elementary steps have terms for reaction intermediates, and such terms cannot appear in the rate law for the overall reaction.

73. a. I + B –> C + D
b. Fast. If the second step were not fast, the rate equations for the overall reaction and the slow first step would not be the same.

75. The reaction profile for the surface-catalyzed reaction is rather complex, but the highest energy point on the profile is still considerably below the transition state energy in the reaction profile of the noncatalyzed, homogeneous reaction, meaning that the surface-catalyzed reaction has the lower activation energy.

77. Normal body temperature is 37 °C. The enzyme will be less active at 37 °C and even less active at 40 °C.

79. An inhibitor may block the active site of the enzyme or it may react with the enzyme to change the shape of the active site.

81. $0.00500 \text{ L} \times 0.882 \text{ M } H_2O_2 \times \frac{2 \text{ mol } MnO_4}{5 \text{ mol } H_2O_2} \times \frac{1}{0.0500 \text{ M } MnO_4} \times \frac{1000 \text{ mL}}{1 \text{ L}}$
= 35.3 mL

0 s, 35.3 mL; 60 s, 27.9 mL; 120 s, 22.6 mL; 180 s, 18.3 mL,
240 s, 14.9 mL; 300 s, 11.9 mL; 360 s, 9.44 mL; 420 s, 7.52 mL;
480 s, 6.08 mL; 540 s, 4.80 mL; 600 s, 3.76 mL

83. $k = \frac{0.693}{123 \text{ min}} = 5.63 \times 10^{-3} \text{ min}^{-1}$

$\ln \frac{0.15}{1.00} = -5.63 \times 10^{-3}/\text{min} \times t$

$t = 337$ min

85. a. $$\frac{k_2}{k_1} = \frac{\frac{0.693}{t_2}}{\frac{0.693}{t_1}} = \frac{t_1}{t_2}$$

$$\ln \frac{t_1}{t_2} = \frac{E_a}{R}\left(\frac{1}{T_1} - \frac{1}{T_2}\right)$$

$$\ln \frac{32 \text{ min}}{t_2} = \frac{113 \text{ kJ /mol x } \frac{1000 \text{ J}}{\text{kJ}}}{8.3145 \text{ J/mol K}} \text{ x } \left(\frac{1}{298} - \frac{1}{308}\right) =$$

$$1.36 \text{ x } 10^4 \text{ x } 1.09 \text{ x } 10^{-4} = 1.48$$

$$\frac{32}{t_2} = e^{1.48} = 4.39$$
$$t_2 = 7.29 \text{ min}$$

b. $$\text{rate} = k[\text{PAN}] = \frac{0.693}{7.29 \text{ min}} \text{ x } 6.0 \text{ x } 10^{14} \text{ molecules}$$
$$\text{rate} = 5.7 \text{ x } 10^{13} \text{ molecules/min}$$

87. $$\ln \frac{N_t}{N_o} = -\lambda t \qquad \lambda = \frac{0.693}{5730 \text{ y}} = 1.209 \text{ x } 10^{-4} \text{ y}^{-1}$$
$$\ln \frac{N_t}{15 \text{ dis/min}} = -1.209 \text{ x } 10^{-4} \text{ y}^{-1} \text{ x } 50000 \text{ y} = -6.045$$
$$\frac{N_t}{15 \text{ dis/min}} = e^{-6.045} = 2.37 \text{ x } 10^{-3}$$
$$N_t = 0.036 \text{ dis/min}$$

89. $$1.00 \text{ g Co x } \frac{\text{mol Co}}{60.0 \text{ g}} \text{ x } \frac{6.022 \text{ x } 10^{23} \text{ atoms}}{\text{mol Co}} \text{ x } \frac{0.693}{5.271 \text{ y}} = 1.32 \text{ x } 10^{21} \text{ atom/y}$$

$$X \text{ g Sr x } \frac{1 \text{ mol Sr}}{90.0 \text{ g Sr}} \text{ x } \frac{6.022 \text{ x } 10^{23} \text{ atom}}{1 \text{ mol Sr}} \text{ x } \frac{0.693}{28.5 \text{ y}} = 1.32 \text{ x } 10^{21} \text{ atom/y}$$

$$X = \frac{1.32 \text{ x } 10^{21} \text{ x } 28.5 \text{ x } 90.0}{0.693 \text{ x } 6.022 \text{ x } 10^{23}} = 8.11 \text{ g Sr}$$

91. A larger E_a causes a greater rate increase than a small E_a with the same temperature change. A high E_a barrier means that more molecules can get over it with increasing temperature. A low barrier doesn't stop many, so there is little increase. Another way to visualize this is from the graph of ln$k$ vs $1/T$. The slope of the graph ln$k$ vs $1/T$ is $-E_a/R$. The greater the value of E_a, the greater the slope of the straight-line graph and the more rapidly k increases with temperature.

Chapter 16

Chemical Equilibrium

Exercises

16.1 $K_c = \frac{[COCl_2]}{[CO][Cl_2]} = \frac{[COCl_2]}{[CO]^2}$

There would not be just one value for $[COCl_2]$. There would be a different value of $[COCl_2]$ for each value of CO.

16.2 $K_c = \left(\frac{1}{K_c}\right)^2 = \left(\frac{1}{20}\right)^2 = 2.5 \times 10^{-3}$

16.3 $K_c = \left(\frac{1}{1.8 \times 10^{-6}}\right)^{\frac{1}{2}}$ for $NO_2 \rightleftharpoons NO + \frac{1}{2} O_2$

$K_c = 7.5 \times 10^2$

$K_p = K_c (RT)^{\Delta n} = 7.5 \times 10^2 \times (0.08206 \times 457)^{1/2}$

$K_p = 4.6 \times 10^3$

16.4 $K_p = \frac{P_{CO} P_{H_2}}{P_{H_2O}}$

16.5 A value of K greater than 1 (1.2×10^3) means that the forward reaction is favored. Because K is not extremely large, the reaction will not go to the point where the reactant concentrations are essentially zero.

16.6 a. Reaction would go to the right using up N_2 and producing more NH_3. The amount of H_2, however, would be greater than in the original equilibrium. The amount of N_2 would be less than in the original equilibrium.

b. Reaction would go to the left, using up NH_3 and producing more H_2.

c. Reaction would go to the right by consuming N_2 and H_2. The amount of NH_3 would be less than in the original equilibrium.

16.7 There is no change in the equilibrium amount of HI by changing the pressure or volume, because there are the same number of moles of gas on both sides of the equation.

16.8 At low temperature. The reaction is forced to the right to make more heat by its exothermic reaction and, thus, it is more complete at low temperatures.

16.9 $[H_2] = \frac{10.0\text{ g}}{25.0\text{ L}} \times \frac{\text{mol}}{2.016\text{ g}} = 0.1984\text{ M}$

$[H_2S] = \frac{72.6\text{ g}}{25.0\text{ L}} \times \frac{\text{mol}}{34.09\text{ g}} = 0.08519\text{ M}$

	Sb_2S_3 (s) +	3 H_2 (g) ⇌	2 Sb (s) +	3 H_2S (g)
init		0.1984		--
eq		0.1132		0.08519

$[H_2]_{eq} = 0.1984\text{ M} - [H_2S]_{eq} = 0.1984\text{ M} - 0.0852\text{ M} = 0.1132\text{ M}$

$$K_c = \frac{[H_2S]^3}{[H_2]^3} = \frac{(0.0852)^3}{(0.1132)^3}$$

$K_c = 0.426$

$K_p = K_c(RT)^{\Delta n} = K_c(RT)^0 = 0.426$

16.10 $N_2 + O_2 \rightleftharpoons 2\text{ NO}$

Eq. 1.00-X 1.00-X 2X

$$\frac{(2X)^2}{(1.00-X)^2} = 2.1 \times 10^{-3}$$

$$\frac{2X}{1.00-X} = 0.0458$$

$$2X = 0.0458 - 0.0458X$$

$$X = \frac{0.0458}{2.05} = 0.022\text{ moles}$$

$$\text{mole fraction} = \frac{n_{NO}}{n_{total}} = \frac{2 \times 0.022}{2 \times 0.978 + 2 \times 0.022} = 0.022$$

The factor 2 x 0.978 is $n_{N_2} + n_{O_2}$, where $n_{N_2} = n_{O_2}$. The volume does not matter, as it will cancel out in the equilibrium constant expression.

16.11 $N_2 + O_2 \rightleftharpoons 2\text{ NO}$

Eq. 0.78-X 0.21-X 2X

$$\frac{(2X)^2}{(0.78 - X)(0.21 - X)} = 2.1 \times 10^{-3}$$

$$\frac{4X^2}{0.164 - 0.99X + X^2} = 2.1 \times 10^{-3}$$

$$4X^2 = 2.1 \times 10^{-3}(0.164 - 0.99X + X^2)$$

$$4X^2 = 3.44 \times 10^{-4} - 2.08 \times 10^{-3}X + 2.1 \times 10^{-3}X^2$$

$$4X^2 + 2.08 \times 10^{-3}X - 3.44 \times 10^{-4} = 0$$

$$X = \frac{-2.08 \times 10^{-3} \pm \sqrt{(2.08 \times 10^{-3})^2 + 4 \times 4 \times 3.44 \times 10^{-4}}}{2 \times 4}$$

$$X = \frac{-2.08 \times 10^{-3} \pm 7.42 \times 10^{-2}}{8} = 9.0 \times 10^{-3}$$

moles NO = $2 \times 9.0 \times 10^{-3} = 0.018$ mol
The volume does not matter, as it will cancel out in the equilibrium constant expression.

16.12 $H_2 + I_2 \rightleftharpoons 2\,HI$
$0.0100+X \qquad X \qquad 0.100-2X$
The reaction must go to the left as there is no I_2 present initially.

$$\frac{(0.100-2X)^2}{(0.0100+X)X} = 54.3$$
$$0.0100 - 0.400\,X + 4X^2 = 54.3\,X^2 + 0.543\,X$$
$$50.3\,X^2 + 0.943\,X - 0.0100 = 0$$

$$X = \frac{-0.943 \pm \sqrt{(0.943)^2 - 4 \times 50.3 \times (-0.0100)}}{2 \times 50.3}$$

$X = 7.56 \times 10^{-3}$ or -0.26 not valid
H_2 1.76×10^{-2} moles
I_2 7.56×10^{-3} moles
HI $0.1 - 2 \times 7.56 \times 10^{-3} = 0.0849$ moles

16.13 $K_p = 0.108 = P_{NH_3} \times P_{H_2S}$

$P_{NH_3} = P_{H_2S}$

$P_{NH_3} = \sqrt{0.108} = 0.329$ atm

$P_T = P_{NH_3} + P_{H_2S} = 0.329 \text{ atm} + 0.329 \text{ atm} = 0.658$ atm

Conceptual Exercise

16.1 a. The reverse reaction occurs using up CO_2 and making CO and H_2O. There will be more CO and H_2O and less CO_2. There will be more H_2 as not all of the extra is used up.
b. Because both a reactant (H_2O) and a product (CO_2) are added to the equilibrium mixture, we cannot make a qualitative prediction of whether the equilibrium will shift to the left or to the right.
c. Adding more H_2O favors the forward reaction, as does lowering the temperature (the forward reaction is exothermic). The new equilibrium will have more CO_2, and H_2, and less CO than the original equilibrium. The amount of H_2O is in doubt, because it is not known if the combined effects of the two changes will consume more or less than the 1.00 mol H_2O added.

16.2 If $P_{total} = 0.57$ atm $\quad P = 0.57 - 0.492 = 0.08$ atm

$$K_p = \frac{4\,(0.08)^3}{(0.492 - 2 \times 0.08)^2}$$

$K_p = 0.019$ This is close to 0.023.

Review Questions

12. a. $K_c = \frac{[CO]^2}{[CO_2]}$ b. $K_c = \frac{[HI]^2}{[H_2S]}$ c. $K_c = [O_2]$

13. a. $2\,H_2S\,(g) \rightleftharpoons 2\,H_2\,(g) + S_2\,(g)$ $\quad K_c = \frac{[S_2][H_2]^2}{[H_2S]^2}$

b. $CS_2\,(g) + 4\,H_2\,(g) \rightleftharpoons CH_4\,(g) + 2\,H_2S\,(g)$ $\quad K_c = \frac{[CH_4][H_2S]^2}{[CS_2][H_2]^4}$

c. $CO\,(g) + 3\,H_2\,(g) \rightleftharpoons CH_4\,(g) + H_2O\,(g)$ $\quad K_c = \frac{[CH_4][H_2O]}{[CO][H_2]^3}$

14. a. $K_p = \frac{P_{CO_2}P_{H_2}}{P_{CO}P_{H_2O}}$ b. $K_p = \frac{P^2_{NH_3}}{P_{N_2}P^3_{H_2}}$ c. $K_p = P_{NH_3}P_{H_2S}$

15. a. $1/2\,N_2\,(g) + 1/2\,O_2\,(g) \rightleftharpoons NO\,(g)$, $K_p = \frac{P_{NO}}{(P_{N_2})^{1/2}(P_{O_2})^{1/2}}$

b. $1/2\,N_2\,(g) + 3/2\,H_2\,(g) \rightleftharpoons NH_3\,(g)$, $K_p = \frac{P_{NH_3}}{(P_{N_2})^{1/2}(P_{H_2})^{3/2}}$

c. $1/2\,N_2\,(g) + 1/2\,O_2\,(g) + 1/2\,Cl_2\,(g) \rightleftharpoons NOCl\,(g)$,

$$K_p = \frac{P_{NOCl}}{(P_{N_2})^{1/2}\,(P_{O_2})^{1/2}(P_{Cl_2})^{1/2}}$$

16. a. $2\,CO\,(g) + 2\,NO\,(g) \rightleftharpoons N_2\,(g) + 2\,CO_2\,(g)$ $\quad K_c = \frac{[N_2][CO_2]^2}{[CO]^2[NO]^2}$

b. $5\,O_2\,(g) + 4\,NH_3\,(g) \rightleftharpoons 4\,NO\,(g) + 6\,H_2O\,(g)$ $K_c = \frac{[NO]^4[H_2O]^6}{[NH_3]^4[O_2]^5}$

c. $2\,NaHCO_3\,(s) \rightleftharpoons Na_2CO_3\,(s) + H_2O\,(g) + CO_2\,(g)$ $\quad K_c = [H_2O][CO_2]$

17. $K_c\,(a) = \frac{1}{K_c\,(b)}$

18. K_c for the reaction 1/2 $N_2O_4 \rightleftharpoons NO_2$ is greater than that for the reaction $N_2O_4 \rightleftharpoons 2\ NO_2$. This is because the square root of a number smaller than 1 is *larger* than the number. That is, $\sqrt{0.113} = 0.336$.

20. a. adding HCl b. adding O_2
c. lowering the temperature d. increasing the total pressure on the mixture

21. a. Yes, there is a smaller number of moles of gas on the right.
b. No, there are the same number of moles of gases on both sides of the equation, so a pressure change will have no effect.
c. Yes, there is a smaller number of moles of gas on the right.

22. a. The volume has no effect because there are the same number of moles of gas on both sides of the equation.
b. A larger volume will cause the reaction to go to the right to make more moles of gas.

23. Reaction (a) is exothermic. NO (g) will not dissociate as much at high temperature. Reaction (b) is endothermic and proceeds farther to the right with increased temperature. Dissociation of SO_3 (g) occurs to a greater extent at higher temperatures.

Problems

25. a. $K_c = K_p/(RT)^{\Delta n}$

$$K_c = \frac{2.9 \times 10^{-2}}{(0.08206 \times 303)^1} = 1.17 \times 10^{-3}$$

b. $$K_c = \frac{0.275}{(RT)^1} = \frac{0.275}{(0.08206 \times 700\ K)^1} = 4.79 \times 10^{-3}$$

c. $$K_c = \frac{22.5}{(0.08206 \times (395 + 273))^{-1}} = 1.23 \times 10^3$$

27. $$K_c = \left(\frac{1}{16.7}\right)^2 = 3.59 \times 10^{-3}$$

29. $$K_c = \frac{[C]^2}{[A][B]^2} = \frac{(0.55)^2}{(0.025)(0.15)^2} = 538$$

31. $$K_c = \frac{[COCl_2]}{[CO][Cl_2]} = 1.2 \times 10^3$$

$$1.2 \times 10^3 = \frac{[COCl_2]}{\frac{1}{2}[COCl_2]\,\frac{1}{4}[COCl_2]} = \frac{8}{[COCl_2]}$$

$[COCl_2] = 6.67 \times 10^{-3}$M

33. $$K_p = \frac{P^2_{CO}}{P_{CO_2}} = 63$$

$$63 = \frac{(10P_{CO_2})^2}{P_{CO_2}} = 100\,P_{CO_2}$$

$P_{CO_2} = 0.63$ atm
$P_{CO} = 6.3$ atm
$P_{total} = 6.9$ atm

35. $$K_p = \frac{P^3_{H_2S}}{P^3_{H_2}} = \frac{P^3_{H_2S}}{P^3_{H_2S}} = 1$$

37. $$K_c = \frac{[COCl_2]}{[CO][Cl_2]} = 1.2 \times 10^3$$

$$\frac{0.10}{1.5 \times 2.0} = Q_c = 3.33 \times 10^{-2}$$

$Q_c < K_c$ The reaction will go to the right.
More $COCl_2$ will be produced.

39. $$K_p = 23.2 = \frac{P_{CO_2}P_{H_2}}{P_{CO}P_{H_2O}}$$

a. No, K_p would have to equal 1.
b. Yes, it is possible.
c. No, K_p would have to equal 1.
d. Yes, it is possible.

41. $[NO_2] = \frac{1.353\text{ g}}{10.5\text{ L}} \times \frac{\text{mol}}{46.01\text{ g}} = 2.801 \times 10^{-3}\text{ M}$

$[NO] = \frac{0.0960\text{ g}}{10.5\text{ L}} \times \frac{\text{mol}}{30.01\text{ g}} = 3.047 \times 10^{-4}\text{ M}$

$[O_2] = \frac{0.0512\text{ g}}{10.5\text{ L}} \times \frac{\text{mol}}{32.00\text{ g}} = 1.524 \times 10^{-4}\text{ M}$

$$K_c = \frac{[NO]^2[O_2]}{[NO_2]^2} = \frac{(3.047 \times 10^{-4})^2 \times 1.524 \times 10^{-4}}{(2.801 \times 10^{-3})^2} = 1.80 \times 10^{-6}$$

$K_p = K_c\,(RT)^1 = 1.80 \times 10^{-6} \times (0.08206 \times (184 + 273))$
$K_p = 6.76 \times 10^{-5}$

43. $[H_2] = \frac{0.403\text{ g}}{0.750\text{ L}} \times \frac{\text{mol}}{2.016\text{ g}} = 0.2665\text{ M}$

$[I_2] = \frac{25.4\text{ g}}{0.750\text{ L}} \times \frac{\text{mol}}{253.80\text{ g}} = 0.1334\text{ M}$

$[I_2]_{eq} = \frac{1.40\text{ g}}{0.750\text{ L}} \times \frac{\text{mol}}{253.80\text{ g}} = 0.007355\text{ M}$

	H_2	$+\ I_2$	$\rightleftharpoons 2HI$
init	0.2665	0.1334	
eq	$0.2665 - (0.1334 - 7.355 \times 10^{-3})$	7.355×10^{-3}	$2(0.1334 - 7.355 \times 10^{-3})$

$$K_c = \frac{[HI]^2}{[I_2][H_2]} = \frac{(0.2521)^2}{7.355 \times 10^{-3} \times 0.1405} = 61.5$$

45. $P_{total} = P_{NH_3} + P_{CO_2}$

$P_{NH_3} = 2P_{CO_2}$

$P_{total} = 2P_{CO_2} + P_{CO_2} = 0.164\text{ atm}$

$P_{CO_2} = 0.164\text{ atm}/3 = 0.055\text{ atm}$

$P_{NH_3} = 0.110\text{ atm}$

$K_p = P^2_{NH_3}P_{CO_2} = (0.110)^2(0.055) = 6.66 \times 10^{-4}$

47. These dissociation reactions, because they require the breaking of bonds with no new bonds formed, are endothermic. The forward reaction is favored with increasing temperature -- dissociation is more extensive when equilibrium is reached.

49. a. $N_2 + O_2 \rightleftharpoons 2\ NO$

There is no effect because the same number of moles of gas appear on each side of the equation.

b. $N_2 + 3\ H_2 \rightleftharpoons 2\ NH_3$

Because the number of molecules of gas produced is less than the number of molecules of reactant gases, this reaction occurs to a greater extent with increased pressure.

c. $H_2 + I_2 \rightleftharpoons 2\ HI$

There is no effect because the same number of moles of gas appear on each side of the equation.

d. $2\ H_2 + S_2 \rightleftharpoons 2\ H_2S$

Because fewer molecules of gas are produced, this reaction occurs to a greater extent.

51. The density of ice is less than that of liquid water. An increased pressure on the ice favors the process in which the water molecules occupy a reduced volume -- the liquid state. The ice melts.

53.

C_6H_{12}	$\rightleftharpoons$	$C_5H_9CH_3$
$\frac{100\text{ g}}{84.16\text{ g/mol}} = 1.19$ moles		--
$1.19 - X$		X

$$\frac{X}{1.19 - X} = 0.143$$

$X = 0.170 - 0.143\,X$

$1.143\,X = 0.170$

$X = 0.149$ moles

$0.149 \text{ moles} \times \frac{84.16\text{ g}}{\text{mole}} = 12.5$ g methylcyclopentane

(As long as the same number of concentration terms appear in the numerator and denominator, the volume does not matter.)

55.

	C	+	H_2O	$\rightleftharpoons$	CO	+	H_2
init			0.100		--		0.100
eq			$0.100 - X$		X		$0.100 + X$

$$K_c = 0.111 = \frac{\frac{X}{1.00} \times \frac{0.100 + X}{1.00}}{\frac{0.100 - X}{1.00}} = \frac{X\,(0.100 + X)}{(0.100 - X)}$$

$$0.111 = \frac{0.100\,X + X^2}{0.100 - X}$$

$0.0111 - 0.111\,X = 0.100\,X + X^2$

$X^2 + 0.211\,X - 0.0111 = 0$

$$X = \frac{-0.211 \pm \sqrt{(0.211)^2 - 4 \times (-0.0111)}}{2}$$

$X = 0.0436$ or -0.254 not valid

$X = 0.0436$ mol CO

57.

	$2COF_2$	$\rightleftharpoons$ CO_2	+	CF_4
init	0.500 mol	--		--
eq	X	$0.250 - \frac{1}{2}X$		$0.250 - \frac{1}{2}X$

$$\frac{(0.250 - \frac{1}{2}X)^2}{X^2} = 2.00$$

$$\frac{0.250 - 0.500\,X}{X} = 1.414$$

$1.914\,X = 0.250$

$X = 0.131$ moles COF_2

59. $[PCl_3] = [Cl_2] = [PCl_5] = \frac{0.100 \text{ mol}}{6.40 \text{ L}} = 0.01563$ M

$Q = \frac{0.01563}{(0.01563)^2} = 64$ A net reaction occurs to the left to establish equilibrium.

	PCl_3	+	Cl_2 $\rightleftharpoons$	PCl_5
init	0.01563		0.01563	0.01563
eq	$0.01563 + X$		$0.01563 + X$	$0.01563 - X$

$$\frac{0.01563 - X}{(0.01563 + X)^2} = 26$$

$0.01563 - X = 26\,X^2 + 0.8128\,X + 0.006352$

$0 = 26\,X^2 + 1.8128\,X - 0.009278$

$$X = \frac{-b \pm \sqrt{b^2 - 4ac}}{2a}$$

$$X = \frac{-1.8128 \pm \sqrt{(1.8128)^2 - 4 \times 26 \times (-0.009278)}}{52}$$

$X = 0.004789$ M or -0.07491 M
not valid

PCl_3 and Cl_2 $(0.01563 + 0.00479) \times 6.40$ L = 0.13 mol

PCl_5 $(0.01563 - 0.00479) \times 6.40$ L = 0.069 mol

61. a. $K_P = 0.429 = \frac{P^3_{H_2S}}{P^3_{H_2}} = \frac{(0.200)3}{P^3_{H_2}}$

$P_{H_2} = 0.265$ atm

b. $P_{total} = P_{H_2} + P_{H_2S} = 0.465$ atm

63.

	2 Mo	+ CH_4	⇌	Mo_2C	+ $2H_2$
init		1.00			1.00
eq		1.00 - X			1.00 + 2X

$$K_P = 3.55 = \frac{(1.00 + 2X)^2}{1.00 - X}$$

$$3.55 - 3.55\,X = 1.00 + 4X + 4X^2$$
$$4X^2 + 7.55\,X - 2.55 = 0$$

$$X = \frac{-\,b \pm \sqrt{b^2 - 4ac}}{2a}$$

$$X = \frac{-\,7.55 \pm \sqrt{(7.55)^2 - 4 \times 4 \times (-\,2.55)}}{8}$$

$X = 0.2924$ or -2.18
not valid

$P_{CH_4} = 1.00 - 0.29 = 0.71$ atm
$P_{H_2} = 1.00 + 2 \times 0.29 = 1.58$ atm
$P_{total} = P_{CH_4} + P_{H_2} = 2.29$ atm

65. $$22.44 \text{ mL} \times 0.1025 \text{ M } Ba(OH)_2 \times \frac{2 \text{ mol } CH_3CO_2H}{\text{mol } Ba(OH)_2} \times \frac{10^{-3}\text{ L}}{\text{mL}}$$

$= 4.6002 \times 10^{-3}$ mol CH_3CO_2H

4.6002×10^{-3} mol x 100 = 0.46002 mol CH_3CO_2H

	CH_3CO_2H	+ CH_3CO_2H	⇌	$CH_3CO_2CH_2CH_3$	+ H_2O
init	1.51 mol	1.66 mol		--	--
eq	0.46 mol	1.66 - (1.51 - 0.46)		1.51 - 0.46	1.51 - 0.46

$$K_c = \frac{[CH_3CO_2CH_2CH_3][H_2O]}{[CH_3CO_2H][CH_3CH_2OH]} = \frac{1.05 \times 1.05}{0.46 \times 0.61}$$
$K_c = 3.9$

67. $[N_2O_4] = \frac{0.0508 \text{ mol}}{2.85 \text{ L}} = 1.782 \times 10^{-2}$ M

	N_2O_4	$\rightleftharpoons$	$2\ NO_2$
init	1.782×10^{-2}		--
eq	$1.782 \times 10^{-2} - X$		$2X$

$$K_c = 4.61 \times 10^{-3} = \frac{[NO_2]^2}{[N_2O_4]} = \frac{(2X)^2}{1.782 \times 10^{-2} - X}$$

$$4X^2 = 8.215 \times 10^{-5} - 4.61 \times 10^{-3} X$$

$$4X^2 + 4.61 \times 10^{-3} X - 8.215 \times 10^{-5} = 0$$

$$X = \frac{-b \pm \sqrt{b^2 - 4ac}}{2a}$$

$$X = \frac{-4.61 \times 10^{-3} \pm \sqrt{(4.61 \times 10^{-3})^2 - 4 \times 4 \times (-8.215 \times 10^{-5})}}{8}$$

$X = 3.992 \times 10^{-3}$ M or $-5.145\ 10^{-3}$
not valid

$[N_2O_4] = 1.782 \times 10^{-2} - 3.993 \times 10^{-3} = 1.383 \times 10^{-2}$ M

$[NO_2] = 7.984 \times 10^{-3}$ M

$$P_{N_2O_4} = \frac{nRT}{V} = MRT$$

$$= 1.383 \times 10^{-2} \text{ M} \times \frac{0.08206 \text{ L atm}}{\text{K mol}} \times (273 + 25)\text{K}$$

$= 0.3382$ atm

$$P_{NO_2} = 7.984 \times 10^{-3} \text{ M} \times \frac{0.08206 \text{ L atm}}{\text{K mol}} \times 298 \text{ K}$$

$= 0.1953$ atm

$$P_{total} = P_{N_2O_4} + P_{NO_2} = 0.534 \text{ atm}$$

69. Base the calculation on 100.0 g of the gaseous phase at equilibrium.

$$13.71 \text{ g C} \times \frac{\text{mol}}{12.011 \text{ g}} = 1.1415 \text{ mol C}$$

All the C in the gas phase is present as CS_2.

The number of moles S in CS_2 = 1.142 mol CS_2 x $\frac{2 \text{ mol S}}{1 \text{ mol } CS_2}$ = 2.284 mol S.

$$86.29 \text{ g S} \times \frac{\text{mol}}{32.066 \text{ g}} = 2.691 \text{ mol S}$$

The number of moles S_2 = [2.691 mol S (total) - 2.284 mol S in CS_2] x

$$\frac{1 \text{ mol } S_2}{2 \text{ mol S}} = 0.204 \text{ mol } S_2$$

$$K_c = \frac{n_{CS_2}}{n_{S_2}} = \frac{1.142 \text{ mol } CS_2}{0.204 \text{ mol } S_2} = 5.60$$

71. $P = \frac{nRT}{V} = \frac{0.255 \text{ mol} \times \frac{0.08206 \text{ L atm}}{\text{K mol}} \times 900 \text{ K}}{18.5 \text{ L}}$

$P_{SO_3} = 1.018$ atm

	$2\ SO_3$	$\rightleftharpoons$	$2\ SO_2$	+	O_2
init	1.018 atm		--		--
eq	$1.018 - 2X$		$2X$		X

$K_p = 0.023 = \frac{(2X)^2\ X}{(1.018 - 2X)^2}$

$0.092\ X^2 - 0.0937\ X + 0.0238 = 4X^3$

$4X^3 - 0.092\ X^2 + 0.0937\ X - 0.0238 = 0$

try $X = \sqrt[3]{\frac{0.0238}{4}} = 0.181$

$4 \times (0.181)^3 - 0.092 \times (0.181)^2 + 0.0937 \times (0.181) - 0.0238 = 0.0139$

0.181 is too large.

$4 \times (0.175)^3 - 0.092 \times (0.175)^2 + 0.0937 \times (0.175) - 0.0238 = 0.0112$

0.175 is too large

$4 \times (0.155)^3 - 0.092 \times (0.155)^2 + 0.0937 \times (0.155) - 0.0238 = 03.4 \times 10^{-3}$

0.165 is too small.

$4 \times (0.145)^3 - 0.092 \times (0.145)^2 + 0.0937 \times (0.145) - 0.0238 = 4.67 \times 10^{-5}$

0.145 is still too large, but it is close.

$4 \times (0.14)^3 - 0.092 \times (0.14)^2 + 0.0937 \times (0.14) - 0.0238 = -1.5 \times 10^{-3}$

too small

$X = 0.14$ atm

$P_{total} = P_{SO_3} + P_{SO_2} + P_{O_2}$

$P_{SO_3} = 1.018 - 2\ (0.14) = 0.74$

$P_{SO_2} = 2 \times 0.14 = 0.28$

$P_{O_2} = 0.14$

$P_{total} = 1.16$ atm

73. $P_{O_2} = \chi P^\circ = 0.2095 \times 1.0000 = 0.2095$ atm

$2\ CaSO_4\ (s) \rightleftharpoons 2\ CaO\ (s) + 2\ SO_2\ (g) + O_2\ (g)$

$\qquad X \qquad 0.2095 + \frac{1}{2}X$

$1.45 \times 10^{-5} = P^2_{SO_2} P_{O_2} = X^2 \times (0.2095 + \frac{1}{2}X)$

assume $\frac{1}{2}X << 0.2095$

$1.45 \times 10^{-5} = X^2 \times 0.2095$

$P_{SO_2} = X = 8.32 \times 10^{-3}$ atm

Chapter 17

Acids, Bases, and Acid-Base Equilibria

Exercises

17.1 a. $HS^- + H_2O \rightleftharpoons H_2S + OH^-$
base (1) acid (2) acid (1) base (2)

b. $HNO_3 + H_2PO_4^- \rightleftharpoons H_3PO_4 + NO_3^-$
acid (1) base (2) acid (2) base (1)

17.2 a. H_2Te is the stronger acid. Because the Te atom is larger than the S atom, it is expected that the H-Te bond energy will be less than the H-S bond energy, and the H-Te bond will be more easily broken than the H-S bond.

b. $CH_3CH_2CH_2CHBrCOOH$ is the stronger acid, because the Br on the second carbon is more electron withdrawing than the Cl on the fifth carbon.

17.3 $$[OH^-] = \frac{0.0155 \text{ mol Ba(OH)}_2}{735 \text{ mL}} \times \frac{1000 \text{ mL}}{\text{L}} \times \frac{2 \text{ mol OH}^-}{\text{mol Ba(OH)}_2} = 0.04218 \text{ M}$$

$pOH = -\log [OH^-] = -\log (0.04218) = 1.375$
$pH = 12.625$

17.4 $C_6H_5COOH + H_2O \rightleftharpoons C_6H_5COO^- + H_3O^+$

$$K_a = 6.3 \times 10^{-5} = \frac{[H_3O^+][C_6H_5COO^-]}{[C_6H_5COOH]}$$

$[C_6H_5COOH] = 0.250 - [H_3O^+]$
approximation: $[H_3O^+] = [C_6H_5COO^-]$

$$K_a = 6.3 \times 10^{-5} = \frac{[H_3O^+]^2}{0.250 - [H_3O^+]}$$

assume $0.250 >> [H_3O^+]$
$[H_3O^+] = 4.0 \times 10^{-3}$ M
assumption good
$pH = 2.40$

17.5 $NH_3 + H_2O \rightleftharpoons NH_4^+ + OH^-$

$$K_b = 1.8 \times 10^{-5} = \frac{[OH^-][NH_4^+]}{[NH_3]}$$

approximations: $[OH^-] = [NH_4^+]$, and $[NH_3] = 0.0010\ M - [OH^-]$

$$K_b = 1.8 \times 10^{-5} = \frac{[OH^-]^2}{0.0010\ M - [OH^-]}$$

assume $0.00100 >> [OH^-]$

$[OH^-] = 1.34 \times 10^{-4}$ pOH = 3.87, pH = 10.13

The assumption fails as 1.34×10^{-4} is more than 5% of 0.00100.

$[OH^-]^2 + 1.8 \times 10\text{-}5\ [OH^-] - 1.8 \times 10^{-8} = 0$

$$[OH^-] = \frac{-0\ 1.8 \times 10^{-5} \pm \sqrt{(1.8 \times 10^{-5})^2 - 4 \times 1 \times (-\ 1.8 \times 10^{-8})}}{2}$$

$[OH^-] = 1.25 \times 10^{-4}$ or $-\ 1.43 \times 10^{-4}$ not valid

pOH = 3.90

pH = 10.10

However, to two significant figures, the same result is obtained for $[OH^-]$ by either method: $[OH^-] = 1.3 \times 10^{-4}$ M

17.6 $$K_a = \frac{X^2}{0.0100 - X}$$

$X = 10^{-3.12} = 7.59 \times 10^{-4}$

$$K_a = \frac{(7.59 \times 10^{-4})^2}{0.0100 - 7.59 \times 10^{-4}} = 6.3 \times 10^{-5}$$

$pK_a = 4.20$

17.7 From first ionization, $[H_2PO_4^-] \approx [H_3O^+] = 0.06$ M.

$H_2PO_4^- + H_2O \rightleftharpoons H_3O^+ + HPO_4^{2-}$

$0.06 - X$ $\quad$ $0.06 + X$ $\quad$ X

From second ionization, $\frac{[H_3O^+][HPO_4^{2-}]}{[H_2PO_4^-]} = K_{a_2} = 6.3 \times 10^{-8}$

$$K_{a_2} = 6.3 \times 10^{-8} = \frac{(0.06 + X) \times X}{0.06 - X}$$

assume $0.06 >> X$

$X = 6.3 \times 10^{-8}$ M $= [HPO_4^{2-}]$ $[H_2PO_4^-] = 0.06$ M

17.8 For dilute solutions, reaction goes essentially to completion.

$[H_3O^+] = 2 \times 8.5 \times 10^{-4} = 1.7 \times 10^{-3}$ M

pH = 2.77

17.9 a. $NaNO_3$ is neutral. Na^+ is from the strong base NaOH, NO_3^- is from the strong acid HNO_3.

b. $CH_3CH_2CH_2COOK$ is basic. K^+ is from the strong base KOH. $CH_3CH_2CH_2COO^-$ is from the weak acid.

$$CH_3CH_2CH_2COO^- + H_2O \rightleftharpoons CH_3CH_2CH_2COOH + OH^-$$

17.10 $NH_4^+ + H_2O \rightleftharpoons H_3O^+ + NH_3$

$$K_a = \frac{[H_3O^+][NH_3]}{[NH_4^+]} = \frac{X \cdot X}{0.052 - X} = \frac{K_w}{K_b} = \frac{10^{-14}}{1.8 \times 10^{-5}} = 5.6 \times 10^{-10}$$

assume $X << 0.052$

$X^2 = 0.052 \times 5.6 \times 10^{-10} = 2.9 \times 10^{-11}$

$X = 5.4 \times 10^{-6}$ M

pH = 5.27

17.11 $CH_3COO^- + H_2O \rightleftharpoons CH_3COOH + OH^-$

$$K_b = \frac{K_w}{K_a} = \frac{10^{-14}}{1.8 \times 10^{-5}} = \frac{[OH^-]^2}{[CH_3COO^-]}$$

pH = 9.10 so pOH = 4.90, and $[OH^-] = 10^{-4.90} = 1.26 \times 10^{-5}$

$$\frac{10^{-14}}{1.8 \times 10^{-5}} = \frac{(1.26 \times 10^{-5})^2}{[CH_3COO^-]}$$

$[CH_3COO^-] = 0.29$ M

17.12 $NH_3 \quad + \quad H_2O \rightleftharpoons NH_4^+ \quad + \quad OH^-$

$0.15 - X \qquad\qquad 0.35 + X \qquad X$

$$K_b = 1.8 \times 10^{-5} = \frac{(0.35 + X) \times X}{0.15 - X}$$

assume $X << 0.15$

$$1.8 \times 10^{-5} = \frac{0.35\,X}{0.15}$$

$X = 7.71 \times 10^{-6} = [OH^-]$

pOH = 5.11

pH = 8.89

17.13. $NH_3 \quad + \quad H^+ \rightarrow NH_4^+$

0.24 M $\qquad \frac{0.03}{0.50} = 0.06$ M $\qquad$ 0.20M

0.24 - 0.06 $\qquad\qquad$ 0.20 + 0.06

$$K_b = 1.8 \times 10^{-5} = \frac{0.26}{0.18} \times [OH^-]$$

$[OH^-] = 1.25 \times 10^{-5}$ M

pOH = 4.90

pH = 9.10

or

pK_a for $NH_4^+ = 14.00 - pK_b = 14.00 - (-\log 1.8 \times 10^{-5}) = 9.26$

$$pH = pK_a + \log\frac{[base]}{[conj.\ acid]}$$

$$pH = 9.26 + \log\frac{0.18}{0.26} = 9.10$$

17.14 $[H_3O^+] = \frac{K_w}{K_b}\frac{[NH_4^+]}{[NH_3]}$

$[H_3O^+] = 10^{-9.05} = 8.91 \times 10^{-10}$ M

$[NH_4^+] = \frac{8.91 \times 10^{-10} \times 1.8 \times 10^{-5} \times 0.150}{1 \times 10^{-14}}$

$[NH_4^+] = 0.241$ M

mass = 0.241 M x 0.250 L x 53.49 g/mol = 3.22 g NH_4Cl

An alternate way to work this problem is shown below:

$pH = pK_a + \log\frac{[base]}{[conj.\ acid]}$

$9.05 = 9.26 + \log\frac{0.150}{M}$

$\log\frac{0.150}{M} = -0.21$

$\frac{0.150}{M} = 10^{-0.21} = 0.617$

$M = 0.243$

#g NH_4Cl = 0.243 M x 0.250 L x 53.49 g/mol = 3.25 g NH_4Cl

17.15 $H_3O^+ + OH^- \rightarrow H_2O$

a. $[H_3O^+] = \frac{20.00\ mL \times 0.500\ M - 19.90\ mL \times 0.500\ M}{(20.00 + 19.90)\ mL}$

$[H_3O^+] = 1.253 \times 10^{-3}$

pH = 2.90

b. $[H_3O^+] = \frac{20.00\ mL \times 0.500\ M - 19.99\ mL \times 0.500\ M}{(20.00 + 19.99)\ mL}$

$[H_3O^+] = 1.250 \times 10^{-4}$

pH = 3.90

c. $[OH^-] = \frac{20.01\ mL \times 0.500\ M - 20.00\ mL \times 0.500\ M}{(20.00 + 20.01)\ mL}$

$[OH^-] = 1.250 \times 10^{-4}$

pOH = 3.90

pH = 10.10

d. $[OH^-] = \frac{20.10\ mL \times 0.500\ M - 20.00\ mL \times 0.500\ M}{(20.00 + 20.10)\ mL}$

$[OH^-] = 1.247 \times 10^{-3}$

pOH = 2.90

pH = 11.10

17.16 a. $CH_3COOH +$ OH^- $->$ $H_2O + CH_3COO^-$

     10.00 mmol   12.50 ml x 0.500 M

     <u>-6.25 mmol   -6.25 mmol</u>   <u>+ 6.25 mmol</u>

       3.75 mmol   --   6.25 mmol

This is a buffer solution.

$$[H_3O^+] = Ka\frac{[acid]}{[base]} = 1.8 \times 10^{-5} \times \frac{3.75 \text{ mmol}/32.50 \text{ mL}}{6.25 \text{ mmol}/32.50 \text{ mL}}$$

Notice that since both acid and base are in the same solution, the volume is the same, does cancel out, and can be left out of the calculation.

     $[H_3O^+] = 1.08 \times 10^{-5}$

        pH = 4.97

b. $CH_3COOH +$ OH^- $->$ $H_2O + CH_3COO^-$

     10.00 mmol   20.10 mL x 0.500 M

     <u>-10.00 mmol -10.00 mmol</u>   +10.00 mmol

        0   0.05 mmol

$$[OH^-] = \frac{0.05 \text{ mmol}}{(20.00 + 20.10) \text{ mL}} = 1.25 \times 10^{-3} \text{ M}$$

     pOH = 2.90

      pH = 11.10

Estimation Exercises

17.1 $K_b = \dfrac{[OH^-][NH_4^+]}{[NH_3]} = \dfrac{[OH^-]^2}{[NH_3]}$

$[OH^-] = \sqrt{K_b\,[NH_3]}$ $\sqrt{K_b\,[CH_3NH_2]}$

$[OH^-] = \sqrt{1.8 \times 10^{-5} \times 0.025}$ $\sqrt{4.2 \times 10^{-4} \times 0.030}$

     $= \sqrt{4.5 \times 10^{-7}}$ $\sqrt{1.26 \times 10^{-5}}$

Since both K_b and molarity are larger, it should be obvious that methylamine is more basic.

17.2 Ionization of 0.020 M H_2SO_4 is complete in the first step and partial in the second: 0.020 M < $[H_3O^+]$ < 0.040 M. Only response (b) fits this requirement: $[H_3O^+]$ = 0.25 M. A more exact calculation is below.

$$1.1 \times 10^{-2} = \frac{[H_3O^+][SO_4^{2-}]}{[HSO_4^-]}$$

$$1.1 \times 10^{-2} = \frac{(0.020 + X)\,X}{0.020 - X}$$

$$X^2 + 0.031\,X - 2.2 \times 10^{-4} = 0$$

$$X = \frac{-0.031 \pm \sqrt{(0.031)^2 - 4 \times 1 \times (-2.2 \times 10^{-4})}}{2}$$

$X = 0.0060$ M or -0.037 M
not valid

$[H_3O^+] = 0.020 + 0.006 = 0.026$ M

17.3 The K_a values are $K_a(HNO_2) = 7.2 \times 10^{-4}$ and $K_a(HCN) = 6.2 \times 10^{-10}$. In comparing K_b values for the conjugate ions, that of CN^- is greater than that of NO_2^-, making NH_4CN (aq) more basic than NH_4NO_2 (aq).

Conceptual Exercises

17.1 The solution is basic. The concentration of OH^- comes from the dissociation of NaOH and the self-ionization of water.

17.2
a. NH_4Cl pH = 4.63
b. $NH_4Cl - NH_3$ pH = 9.26
c. $HCl - HNO_3$ pH = - 0.30
d. $CH_3COOH - CH_3COO^-$ pH = 4.74

The indicator color shows that the pH is in the range of about 4 to 5.5. Solutions (b) and (c) have pH values outside this range. The 1.00 M NH_4Cl would have a pH in this range, due to hydrolysis of NH_4^+. The 1.00 M CH_3COOH - CH_3COONa is a buffer with pH = 4.74 (pK_a of acetic acid). To distinguish between (a) and (d), add a small amount of either an acid or a base. The pH of the buffer solution (d) would not change, and that of the 1.00 M NH_4Cl (solution b) would.

17.3 a. At half neutralization $[H_3O^+] = \frac{K_w}{K_b}$

$$10^{-9} = \frac{10^{-14}}{K_b}$$

$K_b = 1.0 \times 10^{-5}$

or

$pH = pK_a \approx 9$

$pK_b = 14.00 - pK_a \approx 5$

$pK_b \approx 1 \times 10^{-5}$

b. From the graph, pH ≈ 5 at equivalence.

At equivalence, hydrolysis of the cation of the weak base occurs.

$$K_a = \frac{K_w}{K_b} = \frac{10^{-14}}{10^{-5}} = \frac{[H_3O^+]^2}{0.50\ M}$$

$[H_3O^+] = 2.24 \times 10^{-5}$

$pH = 4.7$

Review Questions

1. Arrhenius HI (aq) –> $H^+ + I^-$

Brønsted-Lowry HI (aq) + H_2O –> $H_3O^+ + I^-$

4. acid $HPO_4^{2-} + H_2O \rightleftharpoons H_3O^+ + PO_4^{3-}$

base $HPO_4^{2-} + H_2O \rightleftharpoons OH^- + H_2PO_4^-$

5. a. $HClO_2 + H_2O \rightleftharpoons H_3O^+ + ClO_2^-$

b. $CH_3CH_2COOH + H_2O \rightleftharpoons CH_3CH_2COO^- + H_3O^+$

c. $HCN + H_2O \rightleftharpoons H_3O^+ + CN^-$

d. $C_6H_5OH + H_2O \rightleftharpoons H_3O^+ + C_6H_5O^-$

6. a. $K_a = \frac{[H_3O^+][ClO_2^-]}{[HClO_2]}$

b. $K_a = \frac{[H_3O^+][CH_3CH_2COO^-]}{[CH_3CH_2COOH]}$

c. $K_a = \frac{[H_3O^+][CN^-]}{[HCN]}$

d. $K_a = \frac{[H_3O^+][C_6H_5O^-]}{[C_6H_5OH]}$

7. a. $HOClO_2 + H_2O \rightleftharpoons H_3O^+ + OClO_2^-$

acid 1 base 2 acid 2 base 1

b. $HSeO_4^- + NH_3 \rightleftharpoons NH_4^+ + SeO_4^{2-}$

acid 1 base 2 acid 2 base 1

c. $HCO_3^- + OH^- \rightleftharpoons CO_3^{2-} + H_2O$

acid 1 base 2 base 1 acid 2

d. $C_5H_5NH^+ + H_2O \rightleftharpoons C_5H_5N + H_3O^+$

acid 1 base 2 base 1 acid 2

8. (a) benzoic acid < (b) formic acid < (c) hydrofluoric acid < (d) nitrous acid
$K_a = 6.3 \times 10^{-5}$ $\quad 1.8 \times 10^{-4}$ $\quad 6.6 \times 10^{-4}$ $\quad 7.2 \times 10^{-4}$

9. (d) nitrite ion < (c) fluoride ion< (b) formate ion< (a) benzoate ion
$K_b = \frac{K_w}{K_a} = \frac{10^{-14}}{7.2 \times 10^{-4}}$ $\quad \frac{10^{-14}}{6.6 \times 10^{-4}}$ $\quad \frac{10^{-14}}{1.8 \times 10^{-4}}$ $\quad \frac{10^{-14}}{6.3 \times 10^{-5}}$

22. a. strong acid pH < 7.0 b. weak base pH > 7.0
c. weak acid pH < 7.0 d. weak base pH > 7.0

23. a. $KHSO_4$.
HSO_4^- is acidic: $HSO_4^- + H_2O \rightleftharpoons H_3O^+ + SO_4^{2-}$ $K_{a2} = 1.1 \times 10^{-2}$.
K_2SO_4 (aq) should be nearly pH neutral:
$SO_4^{2-} + H_2O \rightleftharpoons OH^- + HSO_4^-$ $K_b = 9 \times 10^{-13}$.
K_2CO_3 (aq) is basic because of the hydroysis of CO_3^{2-}:
$CO_3^{2-} + H_2O \rightleftharpoons HCO_3^- + OH^-$ $K_b = \frac{K_w}{K_{a2}} = 2.1 \times 10^{-4}$
HPO_4^{2-} is only weakly acidic:
$HPO_4^{2-} + H_2O \rightleftharpoons H_3O^+ + PO_4^{3-}$ $K_{a3} = 4.3 \times 10^{-13}$. It is more likely to ionize as a base: $HPO_4^{2-} + H_2O \rightleftharpoons HPO_4^{2-} + OH^-$ $K_b = \frac{K_w}{K_{a2}} = 1.6 \times 10^{-7}$

24. H_2SO_4 < HCl < H_3PO_4 < NH_4NO_3 < KCl < CH_3COONa < NH_3 < KOH

H_2SO_4	HCl	H_3PO_4	NH_4NO_3	KCl	CH_3COONa	NH_3	KOH
diprotic acid strong in 1st ioniz., weak in 2nd	strong monoprotic acid	weak triprotic acid	conjugate acid of weak base	neutral	conjugate base of weak acid	weak base	strong base

Problems

25.

$CH_3COOH + H_2O \rightleftharpoons CH_3COO^- + H_3O^+$
acid (1) base (2) base (1) acid (2)

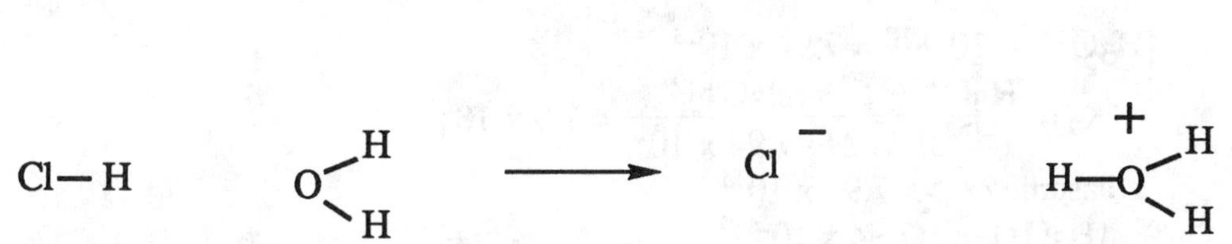

$HCl + H_2O \longrightarrow Cl^- + H_3O^+$
acid (1) base (2) base (1) acid (2)

27. $HCO_3^- + H_3O^+ \rightarrow H_2CO_3 + H_2O \rightarrow 2\ H_2O + CO_2$
It removes H_3O^+.

29. (d) < (b) < (a) < (c)

31. CH_3COOH is a stronger acid than is H_2O. Because aniline is more able to accept a proton from CH_3COOH than from H_2O, it is a stronger base in $CH_3COOH(l)$ than it is in $H_2O(l)$.

33. a. $1.4 \times 10^{-3} < K_a < 8.7 \times 10^{-3}$
Less than 2,2-dichloropropanoic acid but more than 2-chloropropanoic acid.
b. $K_a \approx 3 \times 10^{-5}$ The Cl atom is probably too far from the -COOH group to have much of an electron-withdrawing effect and the K_a value should be only slightly greater than that of unsubstituted acids such as acetic acid ($K_a = 1.8 \times 10^{-5}$) and pentanoic acid ($K_a = 1.5 \times 10^{-5}$).

35.

(e)	<	(a)	<	(c)	<	(b)	<	(d)
~1×10^{-5}		6.3×10^{-5}		3.9×10^{-4}				

37. pOH = 14.00 - pH = 14.00 - 11.13 = 2.87
$[OH^-] = 10^{-2.87} = 1.3 \times 10^{-3}$ M

39. a. $[OH^-] = 2.5 \times 10^{-3}$ M pOH = 2.60
b. $[H_3O^+] = 3.2 \times 10^{-3}$ M pH = 2.49 pOH = 11.51
c. $[OH^-] = 2 \times 3.6 \times 10^{-4}$ M pOH = 3.14
d. $[H_3O^+] = 2.2 \times 10^{-4}$ pH = 3.66 pOH = 10.34

41. $[OH^-] = 2 \times 0.0062 = 0.0124$ M pOH = 1.91 pH = 12.09
The $Ba(OH)_2$ is more basic.

43. $[H_3O^+] = 10^{-1.75} = 1.78 \times 10^{-2}$ M

$$0.0178 \text{ M} \times 0.250 \text{ L} \times \frac{36.45 \text{ g HCl}}{\text{mol HCl}} \times \frac{100 \text{ g solution}}{37.2 \text{ g HCl}} = 0.44 \text{ g conc. HCl}$$

45. $$1.8 \times 10^{-4} = \frac{[H_3O^+][HCOO^-]}{[HCOOH]} = \frac{[H_3O^+]^2}{1.50 \text{ M}}$$
$[H_3O^+] = 1.6 \times 10^{-2}$
pH = 1.78

47. $[H_3O^+] = 10^{-3.10} = 7.94 \times 10^{-4}$

$$K_a = \frac{[H_3O^+]^2}{[HN_3]} = \frac{[H_3O^+]^2}{M - 7.94 \times 10^{-4}} = 1.9 \times 10^{-5}$$

assume $M >> 7.94 \times 10^{-4}$

$$\frac{[H_3O^+]^2}{M} = \frac{(7.94 \times 10^{-4})^2}{M} = 1.9 \times 10^{-5}$$

$M = 3.3 \times 10^{-2}$

49. $[\text{aspirin}] = \frac{1.00\ \text{g}}{0.300\ \text{L}} \times \frac{\text{mol}}{180.15\ \text{g}} = 1.850 \times 10^{-2}$

$[H_3O^+] = 10^{-2.62} = 2.40 \times 10^{-3}$

$$K_a = \frac{[H_3O^+]^2}{M_{aspirin} - [H_3O^+]} = \frac{(2.40 \times 10^{-3})^2}{(1.85 \times 10^{-2} - 2.40 \times 10^{-3})} = 3.6 \times 10^{-4}$$

51. 0.0045 M H_2SO_4 has the lower pH. H_2SO_4 is a strong acid in its first ionization step; moreover, even K_{a_2} of H_2SO_4 is larger than K_{a_1} of H_3PO_4.

53. $\frac{0.05\ \text{g}\ H_3PO_4}{100\ \text{g solution}} \times \frac{1\ \text{g}}{\text{mL}} \times \frac{\text{mL}}{10^{-3}\ \text{L}} \times \frac{\text{mol}}{97.99\text{g}} = 5.1 \times 10^{-3}\ \text{M}$

$$7.1 \times 10^{-3} = \frac{[H_3O^+]^2}{5.1 \times 10^{-3} - [H_3O^+]}$$

$[H_3O^+]^2 + 7.1 \times 10^{-3}\,[H_3O^+] - 3.62 \times 10^{-5} = 0$

$$[H_3O^+] = \frac{-7.10 \times 10^{-3} \pm \sqrt{(7.1 \times 10^{-3})^2 - 4 \times (-3.62 \times 10^{-5})}}{2}$$

$[H_3O^+] = 3.43 \times 10^{-3}$ or -0.011 not valid

pH = 2.5

55. $CH_3CH_2COO^- + H_2O \rightleftharpoons CH_3CH_2COOH + OH^-$

basic

57. $CO_3^{2-} + H_2O \rightleftharpoons HCO_3^- + OH^-$

$$K_b = \frac{K_w}{K_{a2}} = \frac{10^{-14}}{4.7 \times 10^{-11}} = 2.1 \times 10^{-4}$$

59. (c) NH_4I. NH_4^+ is the conjugate acid of the base NH_3, and produces an acidic solution by hydrolysis. Of the other solutions, (a) is neutral and (b) and (d) are basic due to hydrolysis of the anions..

61. $$K_b = \frac{[HOCl][OH^-]}{[OCl^-]} = \frac{[OH^-]^2}{0.080 - [OH^-]} \approx \frac{[OH^-]^2}{0.080} = 3.4 \times 10^{-7}$$

$[OH^-] = 1.65 \times 10^{-4}$

pOH = 3.78

pH = 10.22

63. $[OH^-] = 10^{-(14.00 - 9.05)} = 1.12 \times 10^{-5}$

$CH_3CO_2^- + H_2O \rightleftharpoons CH_3CO_2H + OH^-$

$$K_b = \frac{K_w}{K_a} = \frac{10^{-14}}{1.8 \times 10^{-5}} = \frac{[OH^-]^2}{[CH_3CO_2^-]} = \frac{(1.12 \times 10^{-5})^2}{[CH_3CO_2^-]}$$

$[CH_3CO_2^-] = 0.22$ M

65. $HCOOH + H_2O \rightleftharpoons HCOO^- + H_3O^+$

(c) or (d) The H_3O^+ ion from HNO_3 or the $HCOO^-$ ion from $(HCOO)_2Ca$ will suppress the reaction.

67. $NH_3 + H_2O \rightleftharpoons NH_4^+ + OH^-$

$0.15\text{ M} \qquad\qquad 0.015\text{ M}$

$$K_b = \frac{[NH_4^+][OH^-]}{[NH_3]} = 1.8 \times 10^{-5}$$

$$1.8 \times 10^{-5} = \frac{X\,(0.015 + X)}{0.15 - X}$$

assume that $0.015 >> X$ then

$$1.8 \times 10^{-5} = \frac{X \times 0.015}{0.15}$$

$$X = 1.8 \times 10^{-4} = [NH_4^+]$$

69. $$K_a = \frac{[H_3O^+][CH_3CH_2COO^-]}{[CH_3CH_2COOH]}$$

assume $[H_3O^+] << CH_3CH_2COO^-$

$$1.3 \times 10^{-5} = \frac{X \times 0.0786}{0.350}$$

$$X = 5.79 \times 10^{-5} = [H_3O^+]$$

pH = 4.24

71. a. No, the components need to be a base and its *conjugate* acid or an acid and its *conjugate* base. The HCl and NaOH would neutralize one another, leaving NaCl (aq) with either excess HCl or excess NaOH, not a buffer solution.
b. No, there has to be a concentration of each ion greater than that generated by an equilibrium reaction of one species, if there is to be any significant buffer action. The ratio $[CH_3COO^-]/[CH_3COOH] << 1$, much too small for the solution to act as a buffer.

73. $$12.5\text{ mL} \times \frac{0.719\text{ g}}{\text{mL}} \times \frac{\text{mol}}{59.11\text{ g}} = 0.152\text{ mol } C_3H_7NH_2$$

$$15.00\text{ g} \times \frac{\text{mol}}{95.57\text{ g}} = 0.157\text{ mol } C_3H_7NH_3Cl$$

$$K_b = \frac{[C_3H_7NH_3^+][OH^-]}{[C_3H_7NH_2]} = \frac{[C_3H_7NH_3^+]K_w}{[C_3H_7NH_2][H_3O^+]}$$

$\frac{K_w}{K_a}$ $\quad K_a = \frac{[C_3H_7NH_2][H_3O^+]}{[C_3H_7NH_3]}$

$$[H_3O^+] = \frac{K_w}{K_b} \times \frac{[\text{acid}]}{[\text{base}]} = \frac{K_w}{K_b} \times \frac{[C_3H_7NH_3^+]}{[C_3H_7NH_2]} = \frac{10^{-14}}{3.5 \times 10^{-4}} \times \frac{(0.157)}{(0.152)} = 2.95 \times 10^{-11}$$

pH = 10.53

75.

	$C_3H_7NH_3^+$	+	OH^- ⟶	$C_3H_7NH_2$
initial:	$\frac{0.157\text{ moL}}{0.725\text{ L}} \times 0.0750\text{ L} =$		$0.00200\text{ L} \times 0.0850\text{ M} =$	$\frac{0.152\text{ mol}}{0.725\text{ L}} \times 0.0750\text{ L} =$
	0.0162 mol		0.00017mol	0.0157mol
change:	-0.00017mol		-0.00017mol	+0.00017mol
equil:	0.0160 mol		--	0.0159 mol

$$K_b = \frac{[C_3H_7NH_3^+][OH^-]}{[C_3H_7NH_2]} = \frac{[C_3H_7NH_3^+]K_w}{[C_3H_7NH_2][H_3O^+]}$$

$$[H_3O^+] = \frac{K_w[\text{acid}]}{K_a[\text{base}]} = \frac{K_w[C_3H_7NH_3^+]}{K_a\,[C_3H_7NH_2]} = \frac{10^{-14} \times (0.160)}{3.5 \times 10^{-4} \times (0.159)} = 2.88 \times 10^{-11}$$

pH = 10.55

77. pH = 11.05

$[H_3O^+] = 8.91 \times 10^{-12}$

$$K_b = \frac{[CH_3NH_3^+][OH^-]}{[CH_3NH_2]} = \frac{[CH_3NH_3^+]K_w}{[CH_3NH_2][H_3O^+]}$$

$$[H_3O^+] = \frac{K_w}{K_b}\frac{[CH_3NH_3^+]}{[CH_3NH_2]}$$

$$[CH_3NH_3^+] = \frac{[H_3O^+] \times K_b \times [CH_3NH_2]}{K_w}$$

$$[CH_3NH_3^+] = \frac{8.91 \times 10^{-12} \times 4.2 \times 10^{-4} \times 0.350}{1 \times 10^{-14}}$$

$[CH_3NH_3^+] = 0.131$ M

0.131 M x 0.100 L x 67.52 g/mole CH_3NH_3Cl = 0.88 g CH_3NH_3Cl

79. $$[H_3O^+] = K_a\frac{[\text{acid}]}{[\text{base}]} = 3.98 \times 10^{-9} \times \frac{0.15\text{ M}}{0.10\text{ M}}$$

$[H_3O^+] = 5.97 \times 10^{-9}$

pH = 8.22

81. The new solution will be pH = 7. The new color will be yellow because the phenolphthalein will be colorless, and the thymol blue will be yellow.

83. A test with thymol blue should produce a blue color, and with alizarin yellow R, a yellow color. This places the pH between about 9.6 and 10.0.

85. In contrast to a strong base-strong acid titration, in the titration of a weak base by a strong acid: (1) the initial pH is lower because the weak base is only partially ionized; (2) at the half-neutralization point, pH = pK_b, in a buffer solution in which the concentrations of the weak base and its conjugate acid are equal; (3) the pH < 7 at the equivalence point because the cation of the weak base hydrolyzes; (4) the steep portion of the curve at the equivalence point is confined to a smaller pH range; (5) the choice of indicators is more limited. Only those with a color change in the pH range of about 3 to 6 will work (see Figure 17.16).

87. No. The pH at the equivalence point of the weak base-strong acid titration is well below 7, whereas that of the weak acid-strong base titration is well above 7.

89. $[H_3O^+] = 10^{-2.0} = 1.0 \times 10^{-2}$ M

$$1.0 \times 10^{-2}\text{ M} = \frac{(20.00\text{ mL} \times 0.500\text{ M}) - (X\text{ mL} \times 0.500\text{ M})}{(20.00 + X)\text{ mL}}$$

$0.200 + 0.010\,X = 10.00 - 0.500\,X$

$0.510\,X = 9.80$

$X = 19$ mL

91. 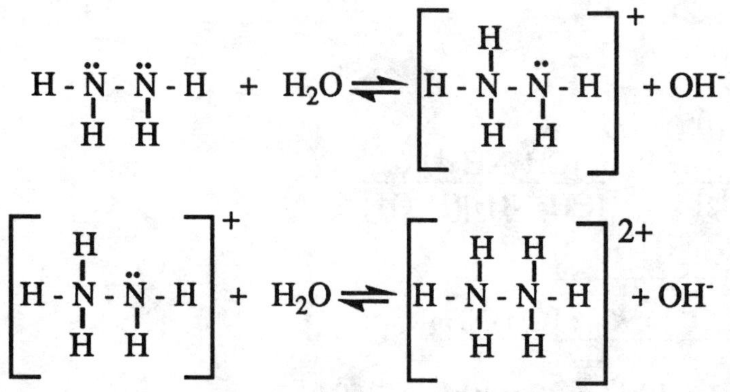

The second H^+ is repelled by the charge on the $NH_3NH_2^+$ ion. This makes $K_{b2} < K_{b1}$, and $pK_{b2} > pK_{b1}$.

93. $CH_3COOH + H_2O \rightleftharpoons CH_3COO^- + H_3O^+$

a. $1.8 \times 10^{-5} = \dfrac{[H_3O^+]^2}{1.0}$

$[H_3O^+] = 4.24 \times 10^{-3}$ M

$\dfrac{4.24 \times 10^{-3}}{1.00} \times 100\% = 0.42\%$

b. $1.8 \times 10^{-5} = \dfrac{[H_3O^+]^2}{0.100}$

$[H_3O^+] = 1.34 \times 10^{-3}$ M

$\dfrac{1.34 \times 10^{-3}}{0.100} \times 100\% = 1.3\%$

c. $1.8 \times 10^{-5} = \dfrac{[H_3O^+]^2}{1.0 \times 10^{-4} - [H_3O^+]}$

$[H_3O^+]^2 + 1.8 \times 10^{-5}[H_3O^+] - 1.8 \times 10^{-9} = 0$

$[H_3O^+] = \dfrac{-1.8 \times 10^{-5} \pm \sqrt{(1.8 \times 10^{-5})^2 - 4 \times 1 \times (-1.8 \times 10^{-9})}}{2}$

$[H_3O^+] = 3.44 \times 10^{-5}$ M or 5.24×10^{-5} not valid

$\dfrac{3.44 \times 10^{-5}}{1.0 \times 10^{-4}} \times 100\% = 34\%$

d. The acidity does not increase. $[H_3O^+]$ *decreases* in the order (a) > (b) > (c). The percent ionization increases with dilution, but it is an increasing percentage of a decreasing total amount of acid.

95. $[H_3O^+] = 10^{-2.20} = 6.31 \times 10^{-3}$ M

$6.31 \times 10^{-3}\text{ M} \times 0.500\text{ L} \times \dfrac{63.02\text{ g}}{\text{mol}} \times \dfrac{100\text{ g solution}}{70.4\text{ g }HNO_3} \times \dfrac{\text{mL}}{1.42\text{ g}} = 0.20\text{ mL}$

97. Yes, as long as $[H_3O^+][OH^-] = K_w$

$$2[OH^-][OH] = 10^{-14}$$
$$[OH^-]^2 = 5 \times 10^{-15}$$
$$[OH^-] = 7.07 \times 10^{-8}$$
$$[H_3O^+] = 1.41 \times 10^{-7}$$

Yes, as long as pH + pOH = 14

$$2pOH + pOH = 14$$
$$3pOH = 14$$
$$pOH = 4.67$$
$$[OH^-] = 2.15 \times 10^{-5}$$
$$pH = 9.33$$
$$[H_3O^+] = 4.67 \times 10^{-10}$$

The solutions having $[H_3O^+] = 2 \times [OH^-]$ and pH = 2 x pOH are not the same. The first is nearly pH neutral, whereas the second is more basic.

99. $K = \frac{c}{p} = \frac{0.033\text{ M}}{1\text{ atm}} = \frac{[H_2CO_3]}{0.00036\text{ atm}}$

$[H_2CO_3] = 1.19 \times 10^{-5}$ M

$$K_{a1} = 4.4 \times 10^{-7} = \frac{[H_3O^+]^2}{1.19 \times 10^{-5} - [H_3O^+]}$$

$$[H_3O^+]^2 + 4.4 \times 10^{-7}[H_3O^+] - 5.2 \times 10^{-12} = 0$$

$$[H_3O^+] = \frac{-4.4 \times 10^{-7} \pm \sqrt{(4.4 \times 10^{-7})^2 - 4 \times 1 (-5.2 \times 10^{-12})}}{2}$$

$[H_3O^+] = 2.1 \times 10^{-6}$ M or -2.45×10^{-6} not valid

pH = 5.68

101. The vertical break is so short and indistinct in a weak base-weak acid titration that it is difficult to determine the equivalence point.

103. The solution at the first equivalence point in the titration of H_3PO_4 (aq) with NaOH (aq) is NaH_2PO_4 (aq). Its pH is represented by the midpoint of the first steep vertical rise in the titration curve. Adding small amounts of acid or base to this solution causes the pH significantly--it is not a buffer solution. A solution containing both $H_2PO_4^-$ and HPO_4^{2-} ions is an effective buffer solution. It is located about in the center of the second very slowly rising portion of the curve.

Chapter 18

Equilibria Involving Slightly Soluble Salts and Complex Ions

Exercises

18.1 a. $Mg(OH)_2\ (s) \rightleftharpoons Mg^{+2} + 2\ OH^-$ $K_{sp} = [Mg^{2+}][OH^-]^2$

b. $Cu_3\ (AsO_4)_2\ (s) \rightleftharpoons 3\ Cu^{2+} + 2\ AsO_4^{3-}$ $K_{sp} = [Cu^{2+}]^3[AsO_4^{3-}]^2$

18.2 $PbF_2 \rightleftharpoons Pb^{2+} + 2\ F^-$

$\quad s \qquad s \qquad 2s$

$$s = \frac{0.064\ g}{100\ mL} \times \frac{mL}{10^{-3}\ L} \times \frac{mol}{245.2\ g} = 2.61 \times 10^{-3}\ M$$

$K_{sp} = (s)(2s)^2 = 4s^3$

$K_{sp} = 4 \times (2.61 \times 10^{-3})^3 = 7.1 \times 10^{-8}$

18.3 $s = 1.7 \times 10^{-4}\ M\ Mg(OH)_2$

$Mg(OH)_2\ (s) \rightleftharpoons Mg^{+2} + 2\ OH^-$

$\quad s \qquad s \qquad 2s$

$K_{sp} = [Mg^{+2}][OH^-]^2 = s\ (2s)^2 = 4s^3$

$K_{sp} = 4 \times (1.7 \times 10^{-4})^3 = 2.0 \times 10^{-11}$

18.4 $Ag_3AsO_4 \rightleftharpoons 3Ag^+ + AsO_4^{3-}$

$\quad s \qquad 3s \qquad s$

$27\ s^4 = 1.0 \times 10^{-22}$

$s = 1.4 \times 10^{-6}\ M$

18.5 $Ag_2SO_4\ (s) \rightleftharpoons 2\ Ag^+ \quad + \quad SO_4^{2-}$

$\quad s \qquad 1.00 + 2s \qquad s$

$K_{sp} = 1.4 \times 10^{-5} = (1.00 + 2s)^2\ s$

assume $1.00 >> 2s$

$1.4 \times 10^{-5} = 1.00\ s$

$s = 1.4 \times 10^{-5}\ M$

18.6 $$[Pb^{2+}] = \frac{1.00\ g}{1.50\ L} \times \frac{mol}{331.2\ g} = 2.013 \times 10^{-3}\ M$$

$$[I^-] = \frac{1.00\ g}{1.50\ L} \times \frac{mol}{278.1\ g} = 2.397 \times 10^{-3}\ M$$

$Q_{ip} = (2.013 \times 10^{-3})(2.397 \times 10^{-3})^2 = 1.16 \times 10^{-8}$

$Q_{ip} > K_{sp}$ A precipitate should form.

18.7 $[I^-] = \dfrac{100 \text{ mL} \times 0.020 \text{ M}}{275 \text{ mL}} = 7.27 \times 10^{-3} \text{ M}$

$[Pb^{2+}] = \dfrac{175 \text{ mL} \times 0.0025 \text{ M}}{275 \text{ mL}} = 1.59 \times 10^{-3} \text{ M}$

$Q_{ip} = [Pb^{2+}][I^-]^2 = (1.59 \times 10^{-3})(7.27 \times 10^{-3})^2$

$Q_{ip} = 8.41 \times 10^{-8}$

$Q_{ip} > K_{sp}$ Precipitation should occur.

18.8

	Ca^{2+} +	$C_2O_4^{2-}$ $\rightleftharpoons$	CaC_2O_4 (s)
Init.	0.0050	0.0100	
Eq.	--	0.0050	0.0050

	CaC_2O_4 (s) $\rightleftharpoons$	Ca^{2+} +	$C_2O_4^{2-}$
Init.		0	0.0050
Eq.		s	0.0050 + s

$K_{sp} = 2.7 \times 10^{-9} = s \times (0.0050 + s)$

assume $s << 0.0050$

$2.7 \times 10^{-9} = s \times 0.0050$

$s = 5.4 \times 10^{-7}$ M

$\dfrac{5.4 \times 10^{-7}}{0.0050} \times 100\% = 0.011\%$

Yes, precipitation is complete.

18.9 $[OH^-] = \left(\dfrac{K_{sp}}{[Fe^{2+}]}\right)^{1/2} = \left(\dfrac{8.0 \times 10^{-16}}{0.10}\right)^{1/2}$

$[OH^-] = 8.9 \times 10^{-8}$

pOH = 7.05

pH = 6.95

18.10 $Ag^+ + 2\, S_2O_3^{2-} \rightleftharpoons [Ag(S_2O_3)_2]^{3-}$

X $1 - 2(0.10 - X)$ $0.10 - X$

$K_f = 1.7 \times 10^{13} = \dfrac{0.10 - X}{X(0.80 + 2X)^2}$

assume $X << 0.10$

$1.7 \times 10^{13} = \dfrac{0.10}{X(0.80)^2}$

$X = [Ag^+] = 9.2 \times 10^{-15}$

18.11 $[I^-] = \dfrac{1.00 \text{ g}}{1.00 \text{ L}} \times \dfrac{\text{mol}}{166.0 \text{ g}} = 6.024 \times 10^{-3}$

$Q_{ip} = [Ag^+][I^-] = 9.2 \times 10^{-15} \times 6.024 \times 10^{-3} = 5.5 \times 10^{-17}$

K_{sp} $8.5 \times 10^{-17} > Q_{ip}$ No, precipitation does not occur.

Estimation Exercises

18.1 The solutes are all of the same type, MX_2. Their molar solubilities parallel their K_{sp} values:

CaF_2	<	PbI_2	<	MgF_2	<	$PbCl_2$
5.3×10^{-9}		7.1×10^{-9}		3.7×10^{-8}		1.6×10^{-5}

18.2 Because $[Ag(CN)_2]^-$ has a larger K_f than the K_f of $[Ag(S_2O_3)_2]^{3-}$ or $[Ag(NH_3)_2]^+$, AgI is most soluble in 0.100 M NaCN.

Conceptual Exercises

18.1
$$Mg(OH)_2\ (s) \rightleftharpoons Mg^{2+}\ (aq) + 2\ OH^-\ (aq)$$
$$2\ NH_4^+\ (aq) + 2\ H_2O\ (l) \rightleftharpoons 2\ NH_3\ (aq) + 2\ H_3O^+\ (aq)$$
$$2\ H_3O^+\ (aq) + 2\ OH^-\ (aq) \rightleftharpoons 4\ H_2O\ (l)$$

$$Mg(OH)_2\ (s) + 2\ NH_4^+\ (aq) \rightleftharpoons Mg^{2+}\ (aq) + 2\ H_2O\ (l) + 2\ NH_3\ (aq)$$

18.2 No. NH_4^+ does not react with NH_3. The complex ion, $[Ag(NH_3)_2]^+$, is not destroyed, and the concentration of free Ag^+ remains too low for AgCl (s) to precipitate.

Review Questions

1. $K_{sp} = [Fe^{3+}][OH^-]^3$

2. $Zn_3(PO_4)_2\ (s) \rightleftharpoons 3\ Zn^{2+}\ (aq) + 2\ PO_4^{2-}\ (aq)$ $\quad K_{sp} = 9.0 \times 10^{-33}$

5. $PbSO_4$ is more soluble because its K_{sp} value is larger and the two solutes are of the same type: MX.

7. $Q_{ip} = [Pb^{2+}][CrO_4^{2-}]$
$= (1.0 \times 10^{-6}\ M)(1.0 \times 10^{-6}\ M) = 1.0 \times 10^{-12}$
$K_{sp} = 2.8 \times 10^{-13}$
$Q_{ip} > K_{sp}$ Yes, precipitation should occur.

11. The solubility of $Fe(OH)_3$ (s) is increased by HCl (aq) and CH_3COOH (aq), both acids, because the H^+ would react with the OH^- from the $Fe(OH)_3$ (s). Its solubility is lowered by $FeCl_3$ (aq), because of the common ion Fe^{3+}; it is also lowered by NaOH (aq) and NH_3 (aq), because of the common ion OH^-.

13. $CH_3CO_2^-\ (aq) + Ag^+\ (aq) \rightleftharpoons AgCH_3CO_2\ (s)$
$H^+\ (aq) + AgCH_3CO_2\ (s) \rightarrow Ag^+\ (aq) + CH_3CO_2H\ (aq)$

15. $Pb(NO_3)_2$ (aq), through the common ion Pb^{2+}, reduces the solubility of $PbCl_2$ (s); but because Pb^{2+} forms the complex ion $[PbCl_3]^-$, HCl (aq) increases the solubility of $PbCl_2$ (s).

16. AgBr (s) dissolves to a greater extent in $Na_2S_2O_3$ (aq) than in NH_3 (aq) because the complex ion $[Ag(S_2O_3)_2]^{3-}$ is more stable than $[Ag(NH_3)_2]^+$.

18. Cation group 1 is separated from other groups by HCl (aq), and Group 2, by H_2S in 0.3 M HCl (aq).

19. $PbCl_2$ (s) is separated from AgCl (s) and Hg_2Cl_2 (s) by hot water. AgCl (s) is separated from Hg_2Cl_2 (s) by NH_3(aq).

Problems

21. a. $K_{sp} = [Fe^{3+}][OH^-]^3$
b. $K_{sp} = [Hg_2^{2+}][Cl^-]^2$
c. $K_{sp} = [Mg^{2+}][NH_4^+][PO_4^{3-}]$
d. $K_{sp} = [Li^+]^3[PO_4^{3-}]$

23. No, the molar solubility and K_{sp} cannot have the same value. The molar solubility must be raised to a power and generally multiplied by a factor to obtain K_{sp}. The molar solubility is larger than K_{sp}, because the solubility is generally much smaller than 1M, and raising such a number to a power (2, 3, 4....) produces a result that is smaller still.

25. $BaCrO_4\ (s) \rightleftharpoons Ba^{2+} + CrO_4^{2-}$
$s \quad s \quad s$
$K_{sp} = s^2$
$$s = \frac{0.0010\ g}{100\ mL} \times \frac{mL}{10^{-3}\ L} \times \frac{mol}{253.33\ g} = 3.95 \times 10^{-5}\ M$$
$K_{sp} = s^2 = 1.6 \times 10^{-9}$

27. $AgCl \rightleftharpoons Ag^+ + Cl^-$
$s \quad s \quad s$
$s^2 = K_{sp} = 1.8 \times 10^{-10}$
$s = 1.3 \times 10^{-5}\ M$
$Ag_2CrO_4 \rightleftharpoons 2\ Ag^+ + CrO_4^{2-}$
$s \quad 2s \quad s$
$4s^3 = K_{sp} = 2.4 \times 10^{-12}$
$s = 8.4 \times 10^{-5}\ M$
Ag_2CrO_4 is more soluble than AgCl.

29.

	$CaCO_3$	$CaSO_4$	CaF_2
K_{sp}	2.8×10^{-9}	9.1×10^{-6}	5.3×10^{-9}

$CaSO_4$ has more $[Ca^{2+}]$ than $CaCO_3$ because of K_{sp} values.
$CaSO_4\ (s) \rightleftharpoons Ca^{2+} + SO_4^{2-}$
$s \quad s \quad s$
$s^2 = 9.1 \times 10^{-6}$
$s = 3.0 \times 10^{-3}\ M$
$CaF_2\ (s) \rightleftharpoons Ca^{2+} + 2\ F^-$
$s \quad s \quad 2s$
$4s^3 = 5.3 \times 10^{-9}$
$s = 1.1 \times 10^{-3}\ M$
$CaSO_4$ also has higher $[Ca^{2+}]$ than CaF_2 but it requires a calculation, not just looking at K_{sp} values.

31. $K_{sp} = 7.6 \times 10^{-36}$

$Cu_3(AsO_4)_2 \rightleftharpoons 3\ Cu^{2+} + 2\ AsO_4^{3-}$

$s \qquad 3s \qquad 2s$

$K_{sp} = 7.6 \times 10^{-36} = (3s)^3(2s)^2$

$7.6 \times 10^{-36} = 108s^5$

$s = 3.71 \times 10^{-8}$ M

$[Cu^{2+}] = 3s = 1.11 \times 10^{-7}$ M

$$\frac{1.11 \times 10^{-7}\ \text{mol}}{\text{L}} \times \frac{10^{-3}\ \text{L}}{\text{mL}} \times \frac{63.55\ \text{g Cu}}{\text{mol Cu}} \times \frac{\text{mL}}{\text{g}} \times \frac{10^6}{10^6} = \frac{7.1\ \text{g Cu}}{10^9\ \text{g}} = 7.1\ \text{ppb}$$

33. AgBr will be most soluble in water, as both of the other solutions contain common ions which reduce solubility.

35. $Mg(OH)_2 \rightleftharpoons Mg^{2+} + 2\ OH^-$

$s \qquad 0.10 + s \qquad 2s$

$K_{sp} = 1.8 \times 10^{-11} = [Mg^{2+}][OH^-]^2 = (0.10 + s)(2s)^2$

assume $0.10 >> s$

$$4s^2 = \frac{1.8 \times 10^{-11}}{0.10} = 1.8 \times 10^{-10}$$

$s = 6.7 \times 10^{-6}$ M

37. $$[I^-] = \frac{15.0\ \text{g}}{0.250\ \text{L}} \times \frac{\text{mol}}{149.89} = 0.400\ \text{M}$$

$PbI_2 \rightleftharpoons Pb^{2+} + 2I^-$

$s \qquad s \qquad 0.400 + 2s$

$K_{sp} = 7.1 \times 10^{-9} = s\,(0.400 + 2s)^2$

assume $0.400 >> 2s$

$s \times (0.400)^2 = 7.1 \times 10^{-9}$

$s = 4.4 \times 10^{-8}\ \text{M} = [Pb^{2+}]$

$[I^-] = 0.400$ M

39. $$[CrO_4^{2-}] = \frac{K_{sp}}{[Ag^+]^2} = \frac{2.4 \times 10^{-12}}{(1.05 \times 10^{-3})^2} = 2.2 \times 10^{-6}$$

41. $$Q_{ip} = [Hg_2^{2-}][Cl^-]^2 = \left(\frac{1.0 \times 10^{-3}}{20}\right)\left(\frac{1.0 \times 10^{-3}}{20}\right)^2$$

$Q_{ip} = 1.25 \times 10^{-13}$

$K_{sp} = 1.3 \times 10^{-18}$ $\qquad Q_{ip} > K_{sp}$ Precipitation occurs.

43. a. $CaSO_4\ (s) \rightleftharpoons Ca^{2+} + SO_4^{2-}$

$s \qquad s \qquad s$

$s^2 = K_{sp} = 9.1 \times 10^{-6}$

$[SO_4^{2-}] = s = 3.0 \times 10^{-3}$ M

b. $$[Pb^{2+}] = \frac{K_{sp}}{[SO_4^{2-}]} = \frac{1.6 \times 10^{-8}}{3.0 \times 10^{-3}} = 5.3 \times 10^{-6}\ \text{M}$$

45. $[OH^-] = 10^{-(14.0 - 9.2)} = 1.6 \times 10^{-5}$

$[Mg^{2+}] = \left(\frac{K_{sp}}{[OH^-]}\right)^2 = \left(\frac{1.8 \times 10^{-11}}{(1.6 \times 10^{-5})}\right)^2 = 7.0 \times 10^{-2}$

$\frac{7.0 \times 10^{-2}}{0.105} \times 100\% = 67\%$

No, precipitation is not complete.

47. $[OH^-] = 10^{-(14.0 - 4.5)} = 3.2 \times 10^{-10}$

$Al(OH)_3 \rightleftharpoons Al^{3+} + 3\ OH^-$

$s \quad s \quad 3.2 \times 10^{-10}$

In a buffer solution, the $[OH^-]$ is constant.

$K_{sp} = 1.3 \times 10^{-33} = s\,(3.2 \times 10^{-10})^3$

$s = 4.0 \times 10^{-5}$ M

49. $NaHSO_4$ is acidic. The H^+ will react to form HCO_3^- and increase the molar solubility.

51. $CaCO_3 + 2\ H^+ \rightarrow Ca^{2+} + H_2CO_3 \rightarrow Ca^{2+} + H_2O + CO_2$

$CaCO_3 + 2\ CH_3COOH$ (vinegar) $\rightarrow Ca^{2+} + 2\ CH_3COO^- + H_2O + CO_2$ (g)

53. $K_3[Fe(CN)_6]$

55. Both H_3O^+ from HCl (d) and HSO_4^- from $NaHSO_4$ (b) can donate protons to NH_3 in the complex ion, causing $[Zn(NH_3)_4]^{2+}$ to dissociate and the concentration of free Zn^{2+} to increase.

57. The H^+ from HNO_3 reacts with the NH_3 leaving Ag^+ and Cl^- to precipitate. When HNO_3 is added to the chloride complex, $[AgCl_2]^-$, the excess Cl^- is not removed, because HCl is a strong acid.

59. $K_f = 1.6 \times 10^7 = \frac{0.15}{1.0 \times 10^{-8}\,[NH_3]^2}$

$[NH_3] = 0.97$ M

61. a. $Zn^{2+} + 4\ OH^- \rightleftharpoons [Zn(OH)_4]^{2-}$

b. Al_2O_3 (s) + 3 H_2O (l) + 6 H_3O^+ (aq) $\rightarrow$ 2 $[Al(H_2O)_6]^{3+}$ (aq)

c. $Fe(OH)_3$ (s) + OH^- $\rightarrow$ N.R.

63. $PbCl_2$ is soluble enough that $[Pb^{2+}]$ remaining in solution after the Group 1 precipitation is sufficiently high that K_{sp} of PbS is exceeded in Group 2. AgCl is so insoluble that $[Ag^+]$ remaining in solution after the Group 1 precipitation is not enough to yield a detectable precipitate of Ag_2S in Group 2.

65. a. Pb^{2+} (aq) + 2 Cl^- (aq) $\rightleftharpoons$ $PbCl_2$ (s)

b. AgCl (s) + Hg_2Cl_2 (s) + 4 NH_3 (aq) $\rightarrow$ $[Ag(NH_3)_2]^+$ (aq) + Hg (l) + $HgNH_2Cl$ (s) + NH_4^+ (aq) + 2 Cl^- (aq)

c. Al^{3+} (aq) + Fe^{3+} (aq) + 7 OH^- (aq) $\rightarrow$ $[Al(OH)_4]^-$ (aq) + $Fe(OH)_3$ (s)

67. Only Hg_2^{2+} is proven to be present, based on the gray color produced when the Group 1 precipitate is treated with NH_3 (aq). The presence of Pb^{2+} and Ag^+ remains uncertain. Treatment of the Group 1 precipitate with hot water and a subsequent test for Pb^{2+} was not performed, and the NH_3 (aq) was not tested for the presence of Ag^+.

69. a. $BaSO_4$ is insoluble enough that so little enough dissolves that it is not dangerous.

b. $BaSO_4 \rightleftharpoons Ba^{2+} + SO_4^{2-}$

$s \quad s \quad s$

$K_{sp} = 1.1 \times 10^{-10} = s^2$

$$s = 1.05 \times 10^{-5}\text{ M} \times \frac{137.33\text{ g}}{\text{mol}} \times \frac{\text{mg}}{10^{-3}\text{ g}} = 1.4\text{ mg/L } Ba^{2+}$$

c. It reduces the solubility of $BaSO_4$.

71. $PbCl_2 \rightleftharpoons Pb + 2\,Cl^-$

$s \quad s \quad 2s$

$K_{sp}\,(80\text{ °C}) = 3.3 \times 10^{-3} = 4s^3$

$s = 0.094\text{ M}$

$K_{sp}\,(25\text{ °C}) = 1.6 \times 10^{-5} = 4s^3$

$s = 0.016\text{ M}$

$0.094 - 0.016 = 0.078\text{ M}$

$$\frac{0.078\text{ mmol}}{\text{mL}} \times 1.00\text{ mL} \times \frac{278.1\text{ mg}}{\text{mmol}} = 22\text{ mg}$$

Yes, it is visible.

73. a. $$[Mg^{2+}] = \frac{K_{sp}}{[OH^-]^2} = \frac{1.8 \times 10^{-11}}{(2.0 \times 10^{-3})^2} = 4.5 \times 10^{-6}\text{ M}$$

b. $$\frac{4.5 \times 10^{-6}}{0.059} \times 100\% = 7.6 \times 10^{-3}\%$$

Yes, precipitation is complete.

75.

	$PbCrO_4$	$PbSO_4$
K_{sp}	2.8×10^{-13}	1.6×10^{-8}

$PbCrO_4$ will appear first.

$$[Pb^{2+}] = \frac{K_{sp}}{[C_2O_4^{2-}]} = \frac{2.8 \times 10^{-13}}{0.010} = 2.8 \times 10^{-11}\text{ M}$$

Pb CrO_4 (s) appears when $[Pb^{2+}] = 2.8 \times 10^{-11}$ M

$$[Pb^{2+}] = \frac{K_{sp}}{[SO_4^{2-}]} = \frac{1.6 \times 10^{-8}}{0.010} = 1.6 \times 10^{-6}\text{ M}$$

$PbSO_4$ (s) appears when $[Pb^{2+}] = 1.6 \times 10^{-6}$ M

Just before $PbSO_4$ (s) begins to precipitate, when $[Pb^{2+}] = 1.6 \times 10^{-6}$ M,

$$[CrO_4^{2-}] = \frac{K_{sp}}{[Pb^{2+}]} = \frac{2.8 \times 10^{-13}}{1.6 \times 10^{-6}} = 1.8 \times 10^{-7}\text{ M}$$

$$\frac{1.8 \times 10^{-7}}{0.010} \times 100\% = 1.8 \times 10^{-3}\%$$

Yes, a separation can be made.

77. $[Cl^-] = \frac{0.100 \text{ mol NaCl}}{0.250 \text{ L}} \times \frac{1 \text{ mol } Cl^-}{1 \text{mol NaCl}} = 0.400 \text{ M}$

$[Ag^+] = \frac{K_{sp}}{[Cl^-]} = \frac{1.8 \times 10^{-10}}{0.400} = 4.5 \times 10^{-10} \text{ M}$

$K_f = 1.6 \times 10^7 = \frac{[Ag(NH_3)^2]^+}{[Ag^+][NH_3]^2}$

$Ag^+ + 2NH_3 \rightleftharpoons [Ag(NH_3)^2]^+$

$0.250 \text{ L} \times 0.100 \text{ M} \times \frac{2 \text{ mol } NH_3}{\text{mol } Ag^+} = 5.0 \times 10^{-2}$ mol NH_3 to form $[Ag(NH_3)_2]^+$

Ag^+	+	$2NH_3$	$\rightleftharpoons$	$[Ag(NH_3)^2]^+$
4.5×10^{-10}		X		0.10

$1.6 \times 10^7 = \frac{0.10}{4.5 \times 10^{-10} [NH_3]^2}$

$[NH_3] = 3.73$ M   This is the concentration of free, uncomplexed NH_3.

3.73 M x 0.250 L = 0.93 mol NH_3

Total number of moles of NH_3 required = 0.93 mol as free NH_3+ 0.05 mol as ligands = 0.98 mol NH_3.

79. a. Some Mg^{2+} compounds are soluble in water and $Mg(OH)_2$ would precipitate from NH_3 (aq),so Mg^{2+} is possibly present.

b. Some Cu^{2+} compounds are soluble in water and $Cu(OH)_2$ would precipitate with high pH, but $Cu(OH)_2$ (s) is not white. Because copper(II) compounds are colored, and Cu^{2+} (aq) is blue, Cu^{2+} is absent (white solids and colorless solutions).

c. Some water insoluble Ba^{2+} compounds would dissolve in HCl (e.g., $BaCO_3$) and would precipitate as $BaSO_4$. Ba^{2+} is possibly present.

d. Na^+ compounds would have been soluble in water but would not have precipitated with NH_3, so Na^+ is absent.

e. NH_4^+ compounds would have dissolved in water but would not have precipitated with NH_3, so NH_4^+ is absent.

Chapter 19

Thermodynamics

Exercises

19.1 a. Spontaneous. The molecules in the wood (principally cellulose, a carbohydrate) would eventually oxidize to CO_2 and H_2O. The decay is greatly enhanced by microorganisms.

b. Nonspontaneous. Stirring NaCl (aq) cannot supply the energy input required to dissociate NaCl into its elements.

c. Indeterminate. $CaCO_3$ (s) should decompose on heating, but whether the decomposition is sufficient to produce CO_2 (g) at 1 atm at 650 °C, cannot be determined without more information.

19.2 a. This is a decrease in entropy because gas goes to solid and two moles go to one.

b. This is an increase in entropy because the gas is formed from solid and two moles go to three.

c. There is no prediction here. All compounds are gas. The same number of moles are on both sides of the equation and the molecules have the same total number of atoms.

19.3 $\Delta S° = \Sigma\nu_p \times S°$ products $- \Sigma\nu_r \times S°$ reactants

$\Delta S° =$ 1 mol CO_2 x 213.6 J /mol K + 1 mol H_2 x 130.6 J/mol K - 1 mol CO x 197.6 J /mol K - 1 mol H_2O x 188.7 J /mol K

$\Delta S° = -42.1$ J /K

19.4 a. $\Delta S < 0, \Delta H < 0$, case 2

b. $\Delta S > 0, \Delta H > 0$, case 3

19.5 a. $\Delta G° = \Delta H° - T\Delta S°$

$$\Delta G° = -114.1 \text{ kJ} - 298 \text{ K} \times (-146.2 \text{ J/K}) \times \frac{\text{kJ}}{10^3 \text{ J}}$$

$\Delta G° = -70.5$ kJ

b. $\Delta G° = \Sigma\nu_p \times \Delta G°_f$ products $- \Sigma\nu_r \times \Delta G°_f$ reactants

$\Delta G° =$ 1 mol CCl_4 (l) x (- 65.27 kJ/mol) + 6 mol S x 0 - 1 mol CS_2 x (+ 65.27 kJ /mol) - 2 mol S_2Cl_2 x (- 31.8 kJ /mol)

$\Delta G° = -66.9$ kJ

19.6 $Mg(OH)_2$ (s) + 2 H_3O^+ –> Mg^{2+} + 4 H_2O (l)

$$K_{eq} = \frac{[Mg^{2+}]}{[H_3O^+]^2}$$

19.7 $\Delta G° =$ 2 mol Hg x 0 + 1 mol O_2 x 0 - 2 mol HgO x - 58.56 kJ /mol

$\Delta G° = 117.12$ kJ

$$\ln K_{eq} = \frac{\Delta G°}{-RT} = \frac{117.12 \text{ kJ} \times 10^3 \text{ J/kJ}}{-8.3145 \text{ J/mol K} \times (273.15 + 25.00)\text{K}} = -47.25$$

$e^{\ln K_{eq}} = K_{eq} = e^{-47.25} = 3.02 \times 10^{-21}$

19.8 $\Delta H^\circ_{rxn} = \Sigma\Delta H^\circ_f$ product - $\Sigma\Delta H^\circ_f$ reactant

ΔH°_{rxn} = 1 mol N_2O_4 x 9.16 kJ /mol - 2 mol NO_2 x 33.18 kJ /mol

ΔH°_{rxn} = - 57.20 kJ

$$\ln\frac{K_2}{K_1} = \frac{\Delta H^\circ}{R}\left(\frac{1}{T_1} - \frac{1}{T_2}\right) = \frac{-57.20 \text{ kJ} \times \frac{10^3 \text{ J}}{\text{kJ}}}{8.3145 \text{ J/mol } K}\left(\frac{1}{298 \text{ K}} - \frac{1}{338 \text{ K}}\right)$$

$$\ln\frac{K_2}{6.9} = -2.732$$

$$\frac{K_2}{6.9} = e^{-2.732} = 6.51 \times 10^{-2}$$

$$K_2 = 0.45$$

Estimation Exercise

19.1 $$\frac{87 \text{ J}}{\text{mol K}} = \frac{\frac{49.45 \text{ kJ}}{\text{mol}} \times \frac{10^3 \text{ J}}{\text{kJ}}}{T}$$

T = 568 K - 273°

T = 295 °C

Conceptual Exercises

19.1 The phase diagram (Figure 11.10) shows that at one atm the equilibrium of CO_2 (s) $\rightleftharpoons$ CO_2 (g) occurs at -78.5 °C. This is the temperature at which the line, P = 1 atm, intersects the sublimation curve of CO_2 (s). At higher T, vaporization goes to completion. At lower T, the vaporization does not occur at 1 atm pressure.

19.2 $$\ln\frac{K_2}{1.00} = \frac{178 \text{ kJ} \times 10^3 \text{ J/kJ}}{8.3145 \text{ J/mol K}}\left(\frac{1}{1170 \text{ K}} - \frac{1}{298 \text{ K}}\right)$$

$$\ln\frac{K_2}{1.00} = -53.5$$

$$\frac{K_2}{1.00} = 5.8 \times 10^{-24}$$

$$K_2 = 5.8 \times 10^{-24}$$

$$\Delta G^\circ = -RT \ln K$$

$$\Delta G^\circ = -8.3145 \text{ J/mol K} \times 298 \text{ K} \times \ln 5.8 \times 10^{-24}$$

$$\Delta G^\circ = 1.33 \times 10^5 \text{ J/mol} \times \frac{\text{kJ}}{10^3 \text{ J}} = 133 \text{ kJ}$$

$$\Delta S^\circ = \frac{\Delta G^\circ - \Delta H^\circ}{-T} = \frac{133 \text{ kJ} - 178 \text{ kJ}}{-298 \text{ K}} =$$

$$\Delta S^\circ = 0.15 \text{ kJ/K}$$

This is substantially is the same quantity as in Conceptual Example 19.2, just a different method of calculating it.

Review Questions

3. a. Spontaneous. Microorganisms are present in the milk that lead to its souring; no further intervention is needed.
   b. Nonspontaneous. The extraction of copper metal from copper ores requires a great deal of external intervention.
   c. Spontaneous. The corrosion of iron in moist air cannot be prevented without external intervention.

4. a. Decrease in entropy. A liquid is converted to a solid.
   b. Increase in entropy. A solid is converted to a gas.
   c. Increase in entropy. A liquid combines with oxygen gas to produce an even greater amount of gaseous products.

9. NOF_3 has more atoms than NO_2F, more modes of vibration and a greater entropy.

11. Low temperature. At low temperatures, the ΔH term dominates in the Gibbs equation, and a $\Delta H < 0$ produces a $\Delta G < 0$.

12. No. Vaporization of water will occur spontaneously, but not to produce a vapor at 1 atm (which occurs only at 100 °C).

18. a. $K_{eq} = \dfrac{P^2_{NCl_3}}{P_{N_2}P^3_{Cl_2}}$

   b. $K_{eq} = [Pb^{2+}][I^-]^2$

   c. $K_{eq} = \dfrac{[H_3O^+][HSO_3^-]}{P_{SO_2}}$

   d. $K_{eq} = \dfrac{[OH^-]^2}{[CO_3^{2-}]}$

Problems

21. An arrangement in which all pennies are heads up is much too ordered to be likely. More likely is that about one-half the pennies will be "heads" and about one-half "tails." This is the highest entropy arrangement.

23. a. Negative - Solids become more ordered as the temperature is lowered.
   b. Positive -The entropy of a gas increases when it is heated and its pressure is lowered.
   c. Negative - A large decrease in entropy occurs if a gas has its pressure increased, its temperature lowered, and is then converted to a solid.

25. a. Decrease - A solid is more ordered than the liquid phase from which it is frozen.
   b. Indeterminate. The gases are all diatomic and the same number of moles of gas appear on each side of the equation.
   c. Increase - A liquid decomposes to produce a large amount of gas.
   d. Decrease - The conversion of a gas to an aqueous solution should produce a more ordered state.

27. b. The value of ΔS is larger for (b). Sublimation of a solid produces much more disorder than conversion of one polymorphic form of a solid to another.

29. Statement (b) is true. If the entropy of a system decreases, that of the surroundings must increase by an even greater amount, so that $\Delta S_{universe} > 0$ and the process is spontaneous.

31. It can be calculated from tables of $S°$.
$\Delta S° = 1 \text{ mol CaO} \times 39.75 \text{ J/mol K} + 1 \text{ mol } CO_2 \times 213.6 \text{ J/mol K}$
$- 1 \text{ mol } CaCO_3 \times 88.70 \text{ J/mol K}$
$\Delta S° = 164.7 \text{ J/mol K}$

33. The disintegration (oxidation) of aluminum is a spontaneous process, but because Al_2O_3 forms a protective coating on the exposed surface which slows down further action of O_2 (g) and H_2O, the complete disintegration takes a very long time. Spontaneity does not involve consideration of the speed of the reaction.

35. In Table 19.1, cases 2 and 3 both represent situations in which the forward reaction is spontaneous at certain temperatures and nonspontaneous at others. When the forward reaction is spontaneous, the reverse reaction is nonspontaneous, and vice versa.

37.

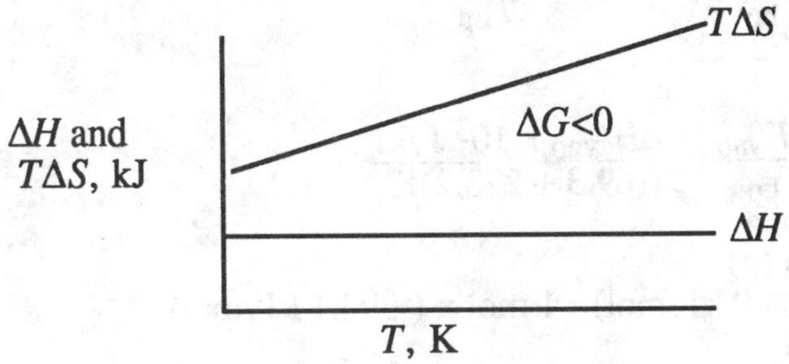

The $T\Delta S$ line does not cross the ΔH line. The $T\Delta S$ line is always above the ΔH line. ΔG is always negative and the reaction is spontaneous.

39. a. Nonspontaneous at all temperatures because $\Delta S < 0$ and $\Delta H > 0$.
b. Spontaneous at low temperatures (because of a large negative value of ΔH). Whether the reaction is spontaneous at very high temperatures is uncertain because the sign of ΔS is difficult to predict--three moles of gaseous reactants produce three moles of gaseous products. (The value of ΔS must be obtained from tabulated data.)
c. Spontaneous at low temperatures but nonspontaneous at high temperatures because $\Delta S < 0$ and $\Delta H < 0$.

41. The reaction should be nonspontaneous at all temperatures because $\Delta S < 0$ and $\Delta H > 0$.

43. a. ΔG° = 1 mol H_2O x (- 228.6 kJ /mol) + 1 mol Fe x 0 - 1 mol FeO x (- 251.5 kJ /mol) - 1 mol H_2 x 0
ΔG° = 22.9 kJ
b. ΔG° = 1 mol H_2O x (- 237.2 kJ /mol) + 1 mol $CdCl_2$ x (-344.0 kJ /mol) - 1 mol CdO x (-228 kJ /mol) - 2 mol HCl x (-95.30 kJ /mol)
ΔG° = - 163 kJ

45. ΔH° = 1 mol CO_2 x (-393.5 kJ /mol) + 2 mol SO_2 x (-296.8 kJ /mol) -1 mol CS_2 x 89.70 kJ /mol - 3 mol O_2 x 0
ΔH° = -1076.8 kJ
ΔS° = 1 mol CO_2 x 213.6 J /mol K + 2 mol SO_2 x (248.1 J /mol K) - 1 mol CS_2 x 151.3 J /mol K - 3 mol O_2 x 205.0 J /mol K
ΔS° = -56.5 J /K
$\Delta G^\circ = \Delta H^\circ - T\Delta S^\circ$
$$\Delta G^\circ = -1076.8 \text{ kJ} - 298 \text{ K} \times (-56.5 \text{ J/K}) \times \frac{\text{kJ}}{10^3 \text{ J}}$$
ΔG° = -1060.0 kJ
ΔG° = 1 mol CO_2 x (-394.4 kJ /mol) + 2 mol SO_2 x (-300.2 kJ /mol) -1 mol CS_2 x 65.27 kJ /mol - 3 mol O_2 x 0 kJ /mol
ΔG° = -1060.1 kJ

47. $$\Delta S^\circ_{vap} = 87 \text{ J/K} = \frac{\Delta H^\circ_{vap}}{T_{bp}} = \frac{30.0 \text{ kJ/mol} \times 10^3 \text{ J/kJ}}{T_{bp}}$$
T_{bp} = 345 K - 273° = 72 °C

49. $$\Delta S^\circ_{vap} = 87 \text{ J/K} = \frac{\Delta H^\circ_{vap}}{T_{bp}} = \frac{\Delta H^\circ_{vap} \times 10^3 \text{ J/kJ}}{(69.3 + 273.2)\text{K}}$$
ΔH°_{vap} = 29.8 kJ /mol
SO_2Cl_2 (l) –> SO_2Cl_2 (g)
ΔH°_{vap} = 1 mol x (-364.0 kJ /mol) - 1 mol x (-394.1 kJ /mol)
ΔH°_{vap} = 30.1 kJ /mol
These values are in good agreement.

51. a. $\Delta G = 0$ at equilibrium
b. $$\ln K_p = \frac{\Delta G^\circ}{-RT} = \frac{-36.3 \text{ kJ} \times 10^3 \text{ J/kJ}}{-8.3145 \text{ J/mol K} \times 850 \text{ K}}$$
$\ln K_p = 5.14$
$K_p = e^{5.14} = 171$

53. a. $K_{eq} = P_{H_2O}P_{SO_2} = K_p$
b. $K_{eq} = [Mg^{2+}][OH^-]^2 = K_{sp}$
c. $$K_{eq} = \frac{[CH_3COOH][OH^-]}{[CH_3COO^-]} = \frac{K_w}{K_a}$$

55. a. $\Delta G° = 2 \text{ mol } SO_3 \text{ x } (-371.1 \text{ kJ /mol}) - 1 \text{ mol } O_2 \text{ x } 0$
$-2 \text{ mol } SO_2 \text{ x } (-300.2 \text{ kJ /mol})$

$\Delta G° = -141.8 \text{ kJ} = -RT \ln K_p$

$$\ln K_p = \frac{-141.8 \text{ kJ /mol x } 10^3 \text{ J /kJ}}{-8.3145 \text{ J /mol K x } 298 \text{ K}} = 57.2$$

$K_p = 7.2 \text{ x } 10^{24}$

b. $\Delta G° = 1 \text{ mol } CO_2 \text{ x } (-394.4 \text{ kJ /mol}) + 4 \text{ mol } H_2 \text{ x } 0$
$- 1 \text{ mol } CH_4 \text{ x } (-50.75 \text{ kJ /mol}) - 2 \text{ mol } H_2O \text{ x } (-228.6 \text{ kJ /mol})$

$\Delta G° = 113.6 \text{ kJ}$

$$\ln K_p = \frac{\Delta G°}{-RT} = \frac{113.6 \text{ kJ x } 10^3 \text{ J /kJ}}{-8.3145 \text{ J /mol K x } 298 \text{ K}}$$

$\ln K_p = -45.8$

$K_p = 1.3 \text{ x } 10^{-20}$

57. $\Delta G° = \Delta H° - T\Delta S°$

$$\Delta G° = 73.6 \text{ kJ /mol} - 298 \text{ K x } 168.7 \text{ J /K mol x } \frac{\text{kJ}}{10^3 \text{ J}}$$

$\Delta G° = 23.3 \text{ kJ /mol}$

$$\ln K_p = \frac{\Delta G°}{-RT} = \frac{23.3 \text{ kJ /mol x } 10^3 \text{ J /kJ}}{-8.3145 \text{ J /mol K x } 298 \text{ K}}$$

$\ln K_p = -9.40$

$K_p = 8.3 \text{ x } 10^{-5} = P_{C_{10}H_8}$

$$P_{C_{10}H_8} = 8.3 \text{ x } 10^{-5} \text{ atm x } \frac{760 \text{ mmHg}}{\text{atm}} = 0.063 \text{ mmHg}$$

59. $$\ln \frac{P_2}{P_1} = \frac{\Delta H°_{vap}}{R}\left(\frac{1}{T_1} - \frac{1}{T_2}\right)$$

$$\ln \frac{P_2}{114 \text{ mmHg}} = \frac{32.5 \text{ kJ /mol x } \frac{10^3 \text{ J}}{\text{kJ}}}{8.3145 \text{ J /mol K}}\left(\frac{1}{298.2 \text{ K}} - \frac{1}{288.2 \text{ K}}\right) = -0.455$$

$$\frac{P_2}{114 \text{ mmHg}} = 0.634$$

$P_2 = 72.3 \text{ mmHg}$

61. $\Delta S^\circ_{vap} = 87\ J/mol = \dfrac{\Delta H^\circ_{vap}}{T_{bp}} = \dfrac{\Delta H^\circ_{vap} \times 10^3\ J/kJ}{(46.6 + 273.2)K}$

$\Delta H^\circ_{vap} = 27.8\ kJ/mol$

$CS_2\ (l) \rightarrow CS_2\ (g)$

$K_p = P_{CS_2}$

$\ln \dfrac{375\ mmHg}{760\ mmHg} = \dfrac{27.8\ kJ/mol \times 10^3\ J/kJ}{8.3145\ J/K\ mol} \times \left(\dfrac{1}{319.8\ K} - \dfrac{1}{T}\right)$

$-0.706 = 3.34 \times 10^3 \times \left(\dfrac{1}{319.8\ K} - \dfrac{1}{T}\right)$

$-2.11 \times 10^{-4} = 3.13 \times 10^{-3} - \dfrac{1}{T}$

$\dfrac{1}{T} = 3.34 \times 10^{-3}$

$T = 300\ K - 273^\circ = 27\ ^\circ C$

63. $\ln \dfrac{2.00}{1.0 \times 10^5} = \dfrac{-41.2\ kJ/mol \times 10^3\ J/kJ}{8.3145\ J/mol\ K} \left(\dfrac{1}{298\ K} - \dfrac{1}{T}\right)$

$2.18 \times 10^{-3} = \left(\dfrac{1}{298\ K} - \dfrac{1}{T}\right)$

$T = 851\ K - 273^\circ = 578\ ^\circ C$

65. $\Delta H^\circ_{vap} = 1\ mol \times (-102.9\ kJ/mol) - 1\ mol \times (-132.2\ kJ/mol)$

$\Delta H^\circ_{vap} = 29.3\ kJ/mol$

$\Delta S^\circ_{vap} = 87\ J/mol\ K = \dfrac{\Delta H^\circ_{vap}}{T_{bp}} = \dfrac{29.3\ kJ/mol \times 10^3\ J/kJ}{T_{bp}}$

$T_{bp} = 337\ K - 273 = 64\ ^\circ C$

67. $\Delta G^\circ_{298} = \Delta G^\circ_f[COCl\ (g)] - \Delta G^\circ_f\ [CO\ (g)] - \Delta G^\circ_f\ [Cl_2\ (g)]$

$\Delta G^\circ_{298} = 1\ mol \times (-206.8\ kJ/mol) - 1\ mol \times (-137.2\ kJ/mol) - 1\ mol \times 0\ kJ/mol = -69.6\ kJ$

$\Delta H^\circ_{298} = \Delta H^\circ_f[COCl\ (g)] - \Delta H^\circ_f\ [CO\ (g)] - \Delta H^\circ_f\ [Cl_2\ (g)]$

$\Delta H^\circ_{298} = 1\ mol \times (-220.9\ kJ/mol) - 1\ mol \times (-110.5\ kJ/mol) - 1\ mol \times 0\ kJ/mol = -110.4\ kJ$

$\Delta G^\circ_{298} = -RT \ln K_p$

$\ln K_p = \dfrac{-\Delta G^\circ_{298}}{RT} = \dfrac{69.6\ kJ/mol \times 10^3\ J/kJ}{8.3145\ J/mol\ K \times 298\ K} = 28.1$

$K_p = e^{28.1} = 1.6 \times 10^{12}$ (at 298 K)

$\ln \dfrac{K_2}{K_1} = \dfrac{\Delta H}{R} \times \left(\dfrac{1}{T_1} - \dfrac{1}{T_2}\right)$

$\ln \dfrac{K_2}{1.6 \times 10^{12}} = \dfrac{-110.4 kJ/mol \times kJ/10^3\ J}{8.3145\ J/mol\ K} \times \left(\dfrac{1}{298} - \dfrac{1}{373}\right) = -8.96$

$\dfrac{K_2}{1.6 \times 10^{12}} = e^{-8.96} = 1.3 \times 10^{-4}$

$K_2 = 2.1 \times 10^8$

69. $\ln\dfrac{57.0\text{ mmHg}}{25.7\text{ mmHg}} = \dfrac{\Delta H^\circ_{vap} \times 10^3\text{ J/kJ}}{8.3145\text{ J/mol K}}\left(\dfrac{1}{308.2\text{ K}} - \dfrac{1}{323.2\text{ K}}\right)$

$$0.797 = \frac{\Delta H^\circ_{vap} \times 10^3\text{ J/kJ}}{8.3145\text{ J/mol K}}(1.506 \times 10^{-4}/\text{K})$$

$$\Delta H^\circ_{vap} = 44.0\text{ kJ/mol}$$

$$\ln\frac{760\text{ mmHg}}{25.7\text{ mmHg}} = \frac{44.0\text{ kJ/mol} \times 10^3\text{ J/kJ}}{8.3145\text{ J/mol K}}\left(\frac{1}{308.2\text{ K}} - \frac{1}{T}\right)$$

$$6.40 \times 10^{-4} = \frac{1}{308.2\text{ K}} - \frac{1}{T} \qquad T = 384\text{ K}$$

$$\Delta S^\circ = \frac{\Delta H^\circ_{vap}}{T_{BP_o}} = \frac{44.0\text{ kJ/mol}}{384\text{ K}} \times \frac{10^3\text{ J}}{\text{kJ}} = 114.6\text{ J/mol K}$$

$$\Delta G^\circ = \Delta H^\circ_{vap} - T\Delta S^\circ_{vap}$$

$$\Delta G^\circ = 44.0\text{ kJ/mol} - 298\text{ K} \times 114.6\text{ J/mol K} \times \frac{\text{kJ}}{10^3\text{ J}}$$

$\Delta G^\circ = 9.8$ kJ /mol
from table $\Delta G^\circ = 159.3$ kJ /mol - 149.2 kJ /mol
$\Delta G^\circ = 10.1$ kJ /mol
These values are close.

71. Energy cannot be created or destroyed, only its form changed. All natural processes have an increase in total entropy.

73. $\Delta G = \Delta G_{\text{urine}} - \Delta G_{\text{blood}}$

$\Delta G = \Delta G°_f - RT \ln c_{\text{urine}} - \Delta G°_f + RT \ln c_{\text{blood}}$

$\Delta G = RT \ln \frac{c_{\text{blood}}}{c_{\text{urine}}} = 8.3145 \text{ J/mol K} \times 310 \text{ K} \times \ln \frac{2.0}{75}$

$\Delta G = -9.34 \times 10^3 \text{ J/mol} \times \frac{\text{kJ}}{10^3 \text{ J}} = -9.34 \text{ kJ/mol}$

75.

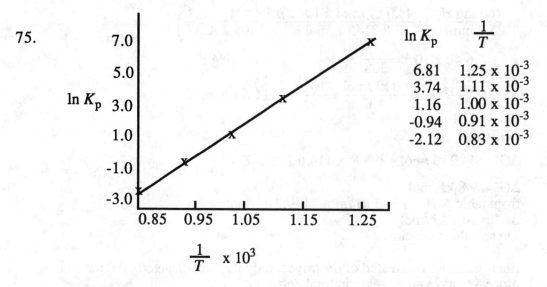

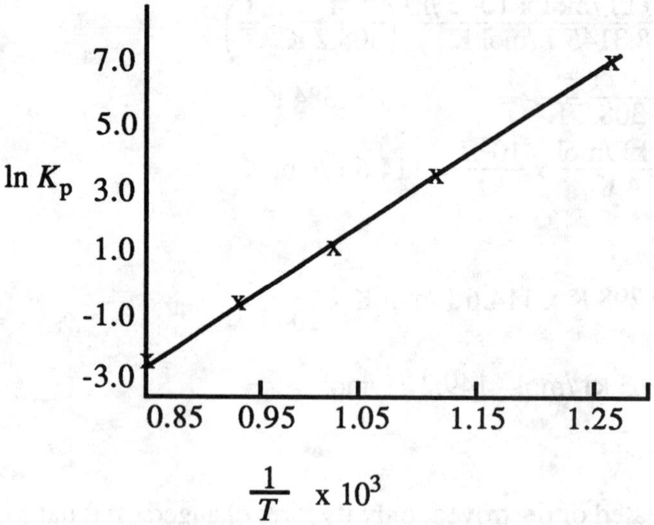

ln K_p	$\frac{1}{T}$
6.81	1.25×10^{-3}
3.74	1.11×10^{-3}
1.16	1.00×10^{-3}
-0.94	0.91×10^{-3}
-2.12	0.83×10^{-3}

$$\ln \frac{K_2}{K_1} = \frac{\Delta H}{R} \times \left(\frac{1}{T_1} - \frac{1}{T_2}\right)$$

$$\frac{\ln K_2 - \ln K_1}{\left(\frac{1}{T_1} - \frac{1}{T_2}\right)} = \frac{\Delta H}{R} = \frac{6.81 - 3.74}{1.11 \times 10^{-3} - 1.25 \times 10^{-3}} = -2.19 \times 10^4 = -\text{ slope}$$

$$\Delta H° = -2.19 \times 10^4 \text{ K} \times 8.314 \text{ J/mol K} \times \frac{\text{kJ}}{10^3 \text{ J}} = -182 \text{ kJ/mol}$$

If two other points on the graph are used:

$$\frac{\ln K_2 - \ln K_1}{\left(\frac{1}{T_1} - \frac{1}{T_2}\right)} = \frac{\Delta H}{R} = \frac{3.74 - 1.16}{1.00 \times 10^{-3} - 1.11 \times 10^{-3}} = -2.35 \times 10^4 = -\text{ slope}$$

$$\Delta H° = -2.35 \times 10^4 \text{ K} \times 8.314 \text{ J/mol K} \times \frac{\text{kJ}}{10^3 \text{ J}} = -195 \text{ kJ/mol.}$$

It appears that the slope is about 2.3×10^4 and $\Delta H° \approx -190$ kJ/mol.
(There are programs to fit the best straight line that would include all of the data points and provide a more accurate answer.)

Chapter 20

Electrochemistry

Exercises

20.1 Anode $\{Al(s) \rightarrow Al^{3+} + 3e^-\} \times 2$

Cathode $\{Cu^{2+} + 2e^- \rightarrow Cu(s)\} \times 3$

$2\,Al(s) + 3\,Cu^{2+}(aq) \rightarrow 3\,Cu(s) + 2\,Al^{3+}(aq)$

20.2 Anode $Zn(s) \rightarrow Zn^{2+}(aq) + 2e^-$

Cathode $Cl_2(g) + 2e^- \rightarrow 2\,Cl^-(aq)$

Net Cell $Zn(s) + Cl_2(g) \rightarrow Zn^{2+}(aq) + 2\,Cl^-(aq)$

$Zn(s)\,|\,Zn^{2+}(aq)\,||\,Cl^-(aq)\,|\,Cl_2(g), Pt$

20.3 $Co^{2+} + 2e^- \rightarrow Co(s)$ $E^\circ_{red} = -0.277$ V

Used in equation as oxidation $E^\circ_{ox} = -(-0.277)$ V

$E^\circ_{cell} = E^\circ_{ox} + E^\circ_{red}$

$E^\circ_{red} = E^\circ_{cell} - E^\circ_{ox} = 1.887 \text{ V} - (0.277 \text{ V})$

$E^\circ_{red} = 1.610$ V

20.4 $Cu^{2+} + 2e^- \rightarrow Cu(s)$ $E^\circ_{red} = 0.337$ V

$\{Fe^{2+} \rightarrow Fe^{3+} + e^-\} \times 2$ $E^\circ_{ox} = -E^\circ_{red} = -0.771$ V

$E^\circ_{cell} = -0.434$ V

Since E°_{cell} is negative, the reaction is not spontaneous in the forward direction.

20.5 $Cu(s) \rightarrow Cu^{2+} + 2e^-$ $E^\circ_{ox} = -E^\circ_{red} = -0.337$ V

$\{Fe^{3+} + e^- \rightarrow Fe^{2+}\} \times 2$ $E^\circ_{red} = 0.771$ V

$E^\circ_{cell} = 0.434$ V

$$E_{cell} = 0.434 \text{ V} = \frac{0.0592}{2} \log K_{eq}$$

$$\log K_{eq} = 14.7$$

$$10^{\log K_{eq}} = K_{eq} = 10^{14.7} = 5 \times 10^{14}$$

20.6 a. $Zn(s) \rightarrow Zn^{2+} + 2e^-$ $E^\circ_{ox} = -E^\circ_{red} = -(-0.763 \text{ V})$

$Cu^{2+} + 2e^- \rightarrow Cu(s)$ $E^\circ_{red} = 0.337$ V

$E^\circ_{cell} = 1.100$ V

$$E_{cell} = E^\circ_{cell} - \frac{0.0592}{n} \log \frac{[Zn^{2+}]}{[Cu^{2+}]}$$

$$E_{cell} = 1.100 \text{ V} - \frac{0.0592}{2} \log \frac{(2.0)}{(0.050)}$$

$$E_{cell} = 1.100 \text{ V} - 0.047 \text{ V}$$

b. $E_{cell} = E°_{cell} - \frac{0.0592}{2} \log \frac{(0.050)}{(2.0)}$

$E_{cell} = 1.100\ V + 0.047\ V$

$E_{cell} = 1.147\ V$

c. $Cu\ (s) \rightarrow Cu^{2+} + 2\ e^-$ $\qquad E°_{ox} = -E°_{red} = -\ 0.337\ V$

$Cl_2\ (g) + 2\ e^- \rightarrow 2\ Cl^-$ $\qquad E°_{red} = 1.358$

$E°_{cell} = 1.021$

$E_{cell} = E°_{cell} - \frac{0.0592}{2} \log \frac{(1.0)(0.25)^2}{(0.50)}$

$E_{cell} = 1.021\ V + 0.027\ V$

$E_{cell} = 1.048\ V$

20.7 $K^+ + e^- \rightarrow K\ (s)$ $\qquad E°_{red} = -\ 2.924\ V$

$2\ H_2O + 2\ e^- \rightarrow H_2\ (g) + 2\ OH^-\ (aq)$ $\qquad E°_{red} = -\ 0.828\ V$

The reduction of H_2O predominates.

$2\ I^- \rightarrow I_2\ (s) + 2\ e^-$ $\qquad E°_{ox} = -E° = -0.535\ V$

$2\ H_2O \rightarrow O_2\ (g) + 4\ H^+ + 4\ e^-$ $\qquad E°_{ox} = -E° = -1.229\ V$

The oxidation of I^- predominates.

$2\ H_2O + 2\ I^- \rightarrow I_2\ (s) + H_2\ (g) + 2\ OH^-\ (aq)$

20.8 $Cu^{2+} + 2\ e^- \rightarrow Cu\ (s)$ $\qquad E°_{red} = 0.337\ V$

$SO_4^{2-}\ (aq) + 4\ H^+\ (aq) + 2\ e^- \rightarrow 2\ H_2O\ (l) + SO_2\ (g)$ $\qquad E°_{red} = 0.17$

The reduction of Cu^{2+} (aq) predominates.

$2\ SO_4^{2-}\ (aq) \rightarrow 2\ e^- + S_2O_8^{2-}\ (aq)$ $\qquad E°_{ox} = -E° = -2.01\ V$

$2\ H_2O\ (l) \rightarrow O_2\ (g) + 4\ H^+ + 4\ e^-$ $\qquad E°_{ox} = -E° = -1.229\ V$

The oxidation of H_2O predominates

$2\ Cu^{2+} + 2\ H_2O \rightarrow 2\ Cu\ (s) + O_2\ (g) + 4\ H^+$

20.9 $Cu^{2+} + 2e^- \rightarrow Cu\ (s)$

$$\frac{1.00\ g\ Cu}{2.25\ A} \times \frac{mol}{63.55\ g} \times \frac{2\ mol\ e^-}{mol\ Cu} \times \frac{96485\ C}{mol\ e^-} \times \frac{1\ A \times 1\ s}{1\ C} \times \frac{min}{60\ s} = 22.5\ min$$

Estimation Exercises

20.1 a. Zn, Al are listed above Co on the table of standard potentials.

b. Co, Fe and Mg are listed above SHE on the table of standard potentials.

20.2 $E = E° - \frac{.0592}{n} \log \frac{[Zn^{2+}]}{[Cu^{2+}]}$

$Zn \rightarrow Zn^{2+} + 2\ e-$ $\qquad -\ (-.763)\ V$

$Cu^{2+} + 2\ e^- \rightarrow Cu$ $\qquad .337\ V$

$Zn\ (s) + Cu^{2+} \rightarrow Zn^{2+} + Cu\ (s)$ $\qquad 1.100\ V$

The larger $[Zn^{2+}]$ and smaller $[Cu^{2+}]$ will make a smaller E. Cell (a) has a 1:1 ratio, $E = 1.100$ V. Cell (c) has a 2:1 ratio and $E = 1.091$ V. This is the lowest value. A small $[Zn^{2+}]$ and large $[Cu^{2+}]$ will produce a negative log term and a larger voltage. Cell (d) (1:10) has a smaller ratio than cell (b) and the largest E_{cell}, 1.130 V.

Conceptual Exercises

20.1 Where the zinc strip touches the citric acid in the lemon, zinc metal atoms give up electrons (oxidation of Zn to Zn^{2+} occurs on the zinc electrode), because zinc is above SHE on the table of standard electrode potentials. The electrons go through the zinc metal, the voltmeter, and the copper metal to where it touches the juice of the lemon. At that point, H^+ is changed to H_2 (g)(the reduction half-reaction is that of H^+(aq) in citric acid to H_2(g)) The lemon juice supplies the H^+ and acts as an electrolyte. (Notice that both strips have to be in the same section of the lemon.) In part this reduction occurs directly on the zinc electrode, but also some electrons pass through the electric measuring circuit to the copper electrode, where reduction of H^+(aq) also occurs

20.2 a. Ag (s), AgI (s) | I^- (1M) anode
Ag (s), AgCl (s) | Cl^- (1M) cathode

b.

AgCl (s) + e^- –> Ag (s) + Cl^-(1 M)	0.222 V
Ag (s) + I^-(1 M) –> e^- + AgI (s)	- (-0.152) V
AgCl (s) + I^- (1 M) –> Cl^- (1 M) + AgI (s)	0.374 V

$E°_{cell}$ = 0.374 V

20.3 Cell A Zn^{2+} + Cu (s) –> Zn (s) + Cu^{2+}
Cell B Zn (s) + Cu^{2+} –> Zn^{2+} + Cu (s)

The current will continue to flow only as long as the concentrations are unequal, that is, there is a force to push the electrons. When all four concentrations are 0.55 M, the current will stop flowing.

Review Questions

3. The anode is the electrode of an electrochemical cell where oxidation occurs; reduction occurs at the cathode.

6. No. The basis of an electrochemical cell reaction must be an oxidation-reduction reaction, and a Brønsted-Lowry acid-base reaction does not involve changes in oxidation states.

9. Standard electrode potentials are based on an arbitrary value of zero assigned to the half-reaction, 2 H^+(1 M) + 2 e^- –> H_2(g, 1 atm). Any reduction half-reaction that has a greater tendency to occur has a positive *E*° value, and any that has a lesser tendency, a negative *E*° value.

11. *E*° is a standard electrode potential that describes the tendency for a particular reduction half-reaction to occur when reactants and products are in their standard states. $E°_{cell}$ is a difference in *E*° values for two half-reactions, and its value depends on the specific half-reactions.

15. Au and Ag have positive reduction potentials and do not react with HCl (aq).

18. $\Delta G^\circ = -nF\,E^\circ_{cell} = -RT \ln K_{eq}$

29. The metal needs to be the cathode, so that metal ions in solution are reduced to metal atoms which become attached to the plated metal.

30. A chlor-alkali process is a process for the electrolysis of NaCl (aq). Chlorine gas ("chlor"), hydrogen gas, and sodium hydroxide solution ("alkaline") are produced.

Problems

31. H_2 (g) –> 2 H^+ + 2 e^- $\quad E^\circ_{ox} = 0.00$ V
Hg_2^{2+} + 2 e^- –> 2 Hg (l) $\quad E^\circ_{red} = ?$
$E^\circ_{cell} = 0.796$ V

$E^\circ_{cell} = E^\circ_{red} + E^\circ_{ox}$
$E^\circ_{red} = E^\circ_{cell} - E^\circ_{ox} = 0.796\text{ V} - 0.00\text{ V} = 0.796\text{ V}$

33. 2 CuI + 2 e^- –> 2 Cu (s) + 2 I^- $\quad E^\circ_{red} = ?$
Cd (s) –> Cd^{2+} + 2 e^- $\quad E^\circ_{ox} = -E^\circ_{red} = -(-0.403\text{ V})$
$E^\circ_{cell} = 0.23$ V

$E^\circ_{cell} = E^\circ_{red} + E^\circ_{ox}$
$0.23\text{ V} = E^\circ_{red} + 0.403\text{ V}$
$E^\circ_{red} = 0.23\text{ V} - 0.403\text{ V} = -0.17\text{ V}$

35. V (s) –> V^{2+} + 2 e^- $\quad E^\circ_{ox} = -E^\circ_{red} = ?$
Cu^{2+} + 2 e^- –> Cu $\quad E^\circ_{red} = 0.337$ V
$E^\circ_{cell} = 1.47$ V

$E^\circ_{cell} = E^\circ_{red} + E^\circ_{ox} = 1.47\text{ V} = E^\circ_{ox} + 0.337\text{ V}$
$E^\circ_{ox} = 1.13\text{ V} = -E^\circ_{red}$
$E^\circ_{red} = -1.13\text{ V}$

37. a. Pb (s) –> Pb^{2+} (aq) + 2 e^- $\quad E^\circ_{ox} = -E^\circ_{red} = -(-0.125\text{ V})$
2 H^+ (aq) + 2 e^- –> H_2 (g) $\quad E^\circ_{red} = 0.00$
Pb (s) + 2 H^+(aq) –> Pb^{2+} (aq) + H_2 (g) $\quad E^\circ_{cell} = 0.125$ V

b. 2 I^- (aq) –> I_2 (s) + 2 e^- $\quad E^\circ_{ox} = -E^\circ_{red} = -(0.535\text{ V})$
Cl_2 (g) + 2 e^- –> 2 Cl^- (aq) $\quad E^\circ_{red} = 1.358$ V
2 I^- (aq) + Cl_2 (g) –> I_2 (s) + 2 Cl^- (aq) $\quad E^\circ_{cell} = 0.823$ V

39. a. Zn (s) –> Zn^{2+} + 2 e^- $\quad E^\circ_{ox} = -E^\circ_{red} = -(-0.763)$ V
{Ag^+ + e^- –> Ag} x 2 $\quad E^\circ_{red} = 0.800$ V
Zn (s) + 2 Ag^+ –> 2 Ag (s) + Zn^{2+} $\quad E^\circ_{cell} = 1.563$ V
Zn (s) | Zn^{2+} (aq) || Ag^+ (aq) | Ag (s)

b. {Fe^{2+} –> Fe^{3+} + e^-} x 4 $\quad E^\circ_{ox} = -E^\circ_{red} = -(0.771\text{ V})$
O_2 (g) + 4 H^+ + 4 e^- –> 2 H_2O $\quad E^\circ_{red} = 1.229$ V
4 Fe^{2+} + O_2 (g) + 4 H^+ –> 4 Fe^{3+} + 2 H_2O $\quad E^\circ_{cell} = 0.458$ V
Pt | Fe^{2+} (aq), Fe^{3+} (aq) || H_2O (l), H^+(aq) | O_2 (g), Pt

41. a. $Sn (s) \rightarrow Sn^{2+} + 2 e^-$ $E°_{ox} = -E°_{red} = -(- 0.137 V)$
$Co^{2+} + 2 e^- \rightarrow Co (s)$ $E°_{red} = - 0.277 V$
not spontaneous $E°_{cell} = - 0.140 V$

b. $2 Br^- \rightarrow Br_2 (l) + 2 e^-$ $E°_{ox} = -E°_{red} = - 1.065 V$
$Cr_2O_7^{2-} + 14 H^+ + 6 e^- \rightarrow 2 Cr^{3+} + 7 H_2O$ $E°_{red} = 1.33 V$
spontaneous $E°_{cell} = 0.26 V$

43. a. $Cd^{2+} + 2 e^- \rightarrow Cd (s)$ $E°_{red} = -0.403 V$
$Al (s) \rightarrow Al^{3+} + 3 e^-$ $E°_{ox} = -E°_{red} = - (-1.676 V)$
spontaneous $E°_{cell} = 1.273 V$

b. $2 Cl^- \rightarrow Cl_2 + 2 e^-$ $E°_{ox} = -E°_{red} = - 1.358 V$
$Br_2 (l) + 2 e^- \rightarrow 2 Br^-$ $E°_{red} = 1.065 V$
not spontaneous $E°_{cell} = - 0.293 V$

c. $Cl^- + 6 OH^- \rightarrow ClO_3^- + 3 H_2O + 6 e^-$ $E°_{ox} = -E°_{red} = - 0.622 V$
$HO_2^- + H_2O + 2 e^- \rightarrow 3 OH^-$ $E°_{red} = 0.88 V$
spontaneous $E°_{cell} = 0.26 V$

45. a. Silver does not react with HCl (aq) because H^+ (aq) is not a good enough oxidizing agent to oxidize Ag (s) to Ag^+ (aq). $E°_{cell}$ for the reaction is - 0.800 V.
b. Nitrate ion in acidic solution is a good enough oxidizing agent to oxidize Ag (s) to Ag+ (aq). $E°_{cell}$ for the reaction is 0.156 V.
$3 Ag (s) + 4 H^+ + NO_3^- \rightarrow 3 Ag^+ + NO (g) + 2 H_2O$

47. Because Cu displaces Rh^{3+} (aq), $E°_{red}$ for the half-reaction $Rh^{3+} + 3 e^- \rightarrow Rh (s)$ must be greater than 0.337 V. Because Ag does not displace Rh^{3+} (aq), $E°_{red}$ for the half-reaction $Rh^{3+} + 3 e^- \rightarrow Rh (s)$ must be less than 0.800 V. Because Rh (s) is oxidized to Rh^{3+} (aq) by HNO_3 (aq), $E°_{ox}$ for $Rh (s) \rightarrow Rh^{3+} + 2 e^-$ must be less negative than -0.956 V. Thus, for the half-reaction $Rh^{3+} + 3 e^- \rightarrow Rh (s)$ $0.337 V < E°_{red} < 0.800 V$.

49. a. $Al (s) \rightarrow Al^{3+} + 3 e^-$ $E°_{ox} = -E°_{red} = -(-1.676 V)$
$\{Ag^+ + e^- \rightarrow Ag (s)\} \times 3$ $E°_{red} = 0.800 V$
$E°_{cell} = 2.476 V$
$\Delta G° = -nFE° = -3 \text{ moles } e^- \times 96485 \text{ C/mol } e^- \times 2.476 V$
$\Delta G° = -716700 J = -716.7 kJ$

b. $\{2 IO_3^- + 12 H^+ + 10 e^- \rightarrow I_2 + 6 H_2O\} \times 2$ $E°_{red} = 1.20 V$
$\{2 H_2O \rightarrow 4 e^- + 4 H^+ + O_2\} \times 5$ $E°_{ox} = -E°_{red} = -1.229 V$
$E°_{cell} = -0.03 V$
$\Delta G° = -nFE° = -20 \text{ mol } e^- \times 96485 \text{ C/mol } e^- \times (-0.03 V)$
$\Delta G° = 5.8 \times 10^4 J = 6 \times 10^1 kJ$

51. a. $Ag^+ + e^- \rightarrow Ag(s)$ $E^\circ_{red} = 0.800$

$Fe^{2+} \rightarrow Fe^{3+} + e^-$ $E^\circ_{ox} = -E^\circ_{red} = -0.771$

$E^\circ_{cell} = 0.029\ V$

$$K_{eq} = \frac{[Fe^{3+}]}{[Fe^{2+}][Ag^+]}$$

$$E^\circ_{cell} = 0.029\ V = \frac{0.0592}{n}\log K_{eq} = \frac{0.0592}{1}\log K_{eq}$$

$$\log K_{eq} = \frac{n\,E^\circ_{cell}}{0.0592} = \frac{1 \times 0.029}{0.0592} = 0.490$$

$$K_{eq} = 10^{0.490} = 3.1$$

b. $MnO_2 + 4\ H^+ + 2\ e^- \rightarrow Mn^{2+} + 2\ H_2O$ $E^\circ_{red} = 1.23\ V$

$2\ Cl^- \rightarrow Cl_2 + 2\ e^-$ $E^\circ_{ox} = -E^\circ_{red} = -1.358\ V$

$E^\circ_{cell} = -0.13\ V$

$$K_{eq} = \frac{[Mn^{2+}]P_{Cl_2}}{[H^+]^4[Cl^-]^2}$$

$$E^\circ_{cell} = -0.13\ V = \frac{0.0592}{n}\log K_{eq} = \frac{0.0592}{2}\log K_{eq}$$

$$\log K_{eq} = \frac{2 \times (-0.13)}{0.0592} = -4.39$$

$$K_{eq} = 10^{-4.39} = 4 \times 10^{-5}$$

c. $\{OCl^- + H_2O + 2\ e^- \rightarrow Cl^- + 2\ OH^-\} \times 2$ $E^\circ_{red} = 0.890$

$4\ OH^- \rightarrow 4\ e^- + 2\ H_2O + O_2$ $E^\circ_{ox} = -E^\circ_{red} = -0.401\ V$

$E^\circ_{cell} = 0.489\ V$

$$K_{eq} = \frac{P_{O_2}[Cl^-]^2}{[OCl^-]^2}$$

$$E^\circ_{cell} = 0.489\ V = \frac{0.0592}{4}\log K_{eq}$$

$$\log K_{eq} = \frac{4 \times 0.489}{0.0592} = 33.04$$

$$K_{eq} = 10^{33.04} = 1 \times 10^{33}$$

53. $Sn \rightarrow Sn^{2+} + 2\ e^-$ $E^\circ_{ox} = -E^\circ_{red} = -(-0.137\ V)$

$Pb^{2+} + 2\ e^- \rightarrow Pb$ $E^\circ_{red} = -0.125\ V$

$E^\circ_{cell} = 0.012\ V$

$$K_{eq} = 10^{\frac{2 \times 0.012}{0.0592}} = 10^{0.405} = 2.5 = \frac{[Sn^{2+}]}{[Pb^{2+}]}$$

	$Sn(s) + Pb^{2+}$	$\rightleftharpoons Sn^{2+} + Pb(s)$
init	1.00 M	--
eq.	$1.00 - X$	X

$$\frac{X}{1.00 - X} = 2.5$$

$X = 2.5 - 2.5X$

$3.5X = 2.5$

$X = 0.71 = [Sn^{2+}]$

$[Pb^{2+}] = 0.29\ M$

55. $$E_{cell} = E°_{cell} - \frac{0.0592}{n} \log \frac{[Zn^{2+}]}{[Pb^{2+}]}$$

a. $$E_{cell} = 0.638 - \frac{0.0592}{2} \log \frac{(0.025)}{(1.55)}$$
$$E_{cell} = 0.691 \text{ V}$$

b. $$E_{cell} = 0.638 - \frac{0.0592}{2} \log \frac{0.50}{0.50}$$
$$E_{cell} = 0.638 \text{ V}$$

c. $$E_{cell} = 0.638 - \frac{0.0592}{2} \log \frac{0.082}{3 \times 10^{-4}}$$
$$E_{cell} = 0.566 \text{ V}$$

57. anode cathode anode cathode

H_2 (g) –> 2 H^+ <=> 2 H^+ + H_2 (g)

$$Q = \frac{[H^+]^2 P_{H_2}}{[H^+]^2 P_{H_2}} \qquad n = 2$$

$$E_{cell} = E°_{cell} - \frac{0.0592}{2} \log \frac{[H^+]^2 (1)}{(1)^2 (1)}$$

$$0.108 = 0 - \frac{0.0592}{2} \log [H^+]^2$$

$- 3.649 = \log [H^+]^2$ or $-3.649 = 2 \log [H^+]$

$[H^+]^2 = 10^{-3.649}$ $-1.825 = \log [H^+]$

$[H^+]^2 = 2.25 \times 10^{-4}$ $[H^+] = 10^{-1.825}$

$[H^+] = 1.5 \times 10^{-2}$ $[H^+] = 1.5 \times 10^{-2}$

pH = 1.82

59. $$E_{cell} = 0.00 - \frac{0.592}{2} \log \frac{[H^+]^2 P_{H_2}}{[H^+]^2 P_{H_2}}$$

H_2 (g) + 2 H^+ <=> 2 H^+ + H_2 (g)

anode cathode anode cathode

$$E_{cell} = 0.00 - \frac{0.0592}{2} \log \frac{(0.0025)^2 (1)}{(1)^2 (1)}$$

$E_{cell} = 0.15$ V

OR

concentration cell: 2 H^+ (1M) –> 2 H^+ (0.0025 M)

$$E_{cell} = - \frac{0.592}{2} \log \frac{[H^+]^2}{[H^+]^2}$$

$$E_{cell} = - \frac{0.0592}{2} \log \frac{(0.0025)^2}{(1)^2}$$

$E_{cell} = 0.15$ V

61. $Zn (s) \rightarrow Zn^{2+} + 2 e^-$ $\qquad E^\circ_{ox} = -E^\circ_{red} = -(-0.763\ V)$
$Cl_2 (g) + 2 e^- \rightarrow 2 Cl^-$ $\qquad E^\circ_{red} = 1.358\ V$

$Zn (s) + Cl_2 (g) \rightarrow Zn^{2+} + 2 Cl^-$ $\qquad E^\circ_{cell} = 2.121\ V$

63. anode: $Zn (s) + 2 OH^- \rightarrow ZnO (s) + H_2O + 2 e^-$
cathode: $Ag_2O + H_2O + 2 e^- \rightarrow 2 Ag + 2 OH^-$

$Zn (s) + Ag_2O (s) \rightarrow ZnO (s) + 2 Ag (s)$

65. Oxygen is the oxidizing agent required to oxidize Fe (s) to Fe^{2+} and then to Fe^{3+}. Water is a reactant in the reduction half-reaction, in which O_2 (g) is reduced to OH^- (aq). Water is also a reactant in the conversion of $Fe(OH)_2$ to $Fe_2O_3 \cdot xH_2O$ ("rust"). The electrolyte completes the electrical circuit between the cathodic and anodic areas.

67. A sacrificial anode is used up, that is, the anode metal is oxidized to metal ions at the same time that a cathodic half-reaction occurs on the protected metal.

69. a. anode $Ni (s) \rightarrow Ni^{2+} + 2 e^-$
cathode $Ni^{2+} + 2 e^- \rightarrow Ni (s)$
Ni (s,anode) –> Ni (s,cathode)
This is similar to the electrorefining of copper.

b. anode $Ni (s) \rightarrow Ni^{2+} + 2 e^-$
cathode $Ni^{2+} + 2 e^- \rightarrow Ni (s)$
Ni (s,anode) –> Ni (s,cathode)

c. anode $2 H_2O \rightarrow 4 H^+ + O_2 + 4 e^-$
cathode $Ni^{2+} + 2 e^- \rightarrow Ni (s)$
$2 H_2O + 2 Ni^{2+} \rightarrow 2 Ni (s) + 4 H^+ + O_2 (g)$

71. a. anode $2 Cl^- \rightarrow Cl_2 + 2 e^-$
cathode $Ba^{2+} + 2 e^- \rightarrow Ba (l)$
$BaCl_2 (l) \xrightarrow{\text{electrolysis}} Ba (l) + Cl_2 (g)$
probable products Ba (l) and Cl_2 (g)

b. anode $2 Br^- \rightarrow Br_2 (l) + 2 e^-$
cathode $2 H^+ + 2 e^- \rightarrow H_2 (g)$
probable products Br_2 (l) and H_2 (g)

c. anode $2 H_2O \rightarrow 4 H^+ + O_2 (g) + 4 e^-$
cathode $\{2 H_2O + 2 e^- \rightarrow H_2 (g) + 2 OH^-\} \times 2$
net $2 H_2O \rightarrow 2 H_2 (g) + O_2 (g)$
probable products O_2 (g) and 2 H_2 (g)

73. $Ag^+ + e^- \rightarrow Ag (s)$

$$1.73\ A \times \frac{C}{A\ s} \times 2.05\ h \times \frac{3600\ s}{h} \times \frac{mol\ e^-}{96485\ C} \times \frac{mol\ Ag}{mol\ e^-} \times \frac{107.9\ g\ Ag}{mol\ Ag} = 14.3\ g\ Ag$$

75. $Cu^{2+} + 2\ e^- \rightarrow Cu$

$$25.0\ g\ Cu \times \frac{mol\ Cu}{63.55\ g\ Cu} \times \frac{2\ mol\ e^-}{mol\ Cu} \times \frac{96485\ C}{mol\ e^-} = 7.59 \times 10^4\ C$$

77. Na (s) does not electrodeposit from $NaNO_3$ (aq). Of the remaining solutions, $AgNO_3$ (aq) yields the greatest number of moles of deposit. One mole of silver is formed for every mole of electrons ($Ag^+ + e^- \rightarrow Ag$), whereas only one-half mole of copper and zinc is formed ($M^{2+} + 2\ e^- \rightarrow M$). Given that the atomic weight of Ag is greater than those of Cu and Zn, $AgNO_3$(aq) (solution c) yields the greatest mass of metal deposit.

$$1.00\ A \times 1\ h \times \frac{3600s}{h} \times \frac{C}{A\ s} \times \frac{mol\ e^-}{96485\ C} \times \frac{mol\ M}{n\ mol\ e} \times \frac{molar\ mass\ g}{mol\ M} =$$

79. Sodium metal is very reactive with water. The sodium will become Na^+ while water is reduced to OH^- and H_2 (g). Instead of just establishing a half-reaction electrode equilibrium, a piece of sodium enters into a complete redox reaction with water.

81. $$1.02\ g\ Ag \times \frac{mol\ Ag}{107.9\ g\ Ag} \times \frac{mol\ e^-}{mol\ Ag} \times \frac{mol\ O_2}{4\ mol\ e^-} \times \frac{22.4\ L}{mol\ O_2} \times \frac{mL}{10^{-3}\ L} = 52.9\ mL\ O_2$$

83. $MnO_2\ (s) + 4\ H^+ + 2\ e^- \rightarrow Mn^{2+} + 2\ H_2O$ $\quad E^\circ_{red} = 1.23\ V$

$2\ Cl^- \rightarrow Cl_2 + 2\ e^-$ $\quad E^\circ_{ox} = -E^\circ_{red} = -1.358\ V$

$MnO_2\ (s) + 4\ H^+ + 2\ Cl^- \rightarrow Cl_2\ (g) + Mn^{2+} + 2\ H_2O$ $\quad E^\circ_{cell} = -0.13\ V$

The Nernst equation will show that, if the concentration of Mn^{2+} is small and the concentrations of Cl^- and H^+ are high, it can make the log term large enough to overcome the negative E°_{cell}. Also, the escape of Cl_2 (g) will push the reaction to the right.

85. $H_2\ (g) \rightarrow 2\ H^+ + 2\ e^-$ $\quad E^\circ_{ox} = -E^\circ_{red} = -0.00\ V$

$2\ H^+ + 2\ e^- \rightarrow H_2\ (g)$ $\quad E^\circ_{red} = 0.00\ V$

$$E_{cell} = E^\circ_{cell} - \frac{0.0592}{2} \log \frac{[H^+]^2}{[H^+]^2}$$

$$E_{cell} = 0.00 - \frac{0.0592}{2} \log \frac{[H^+]^2\ \text{from acid}}{(0.010)^2}$$

$$K_a = 1.8 \times 10^{-5} = \frac{[H^+]^2}{0.45}$$

$$[H^+] = 2.85 \times 10^{-3}$$

$$E_{cell} = 0.00 - \frac{0.0592}{2} \log \frac{(2.85 \times 10^{-3})^2}{(0.010)^2}$$

$$E_{cell} = 0.032\ V$$

87. $Cu\ (s) \rightarrow Cu^{2+} + 2\ e^-$ $\qquad E^\circ_{ox} = -E^\circ_{red} = -0.337\ V$

$Ag^+ + e^- \rightarrow Ag$ $\qquad E^\circ_{red} = 0.800\ V$

$\qquad E^\circ_{cell} = 0.463\ V$

$$\begin{array}{ccc} Ag_2CrO_4 \rightarrow & 2\ Ag^+ + & CrO_4^{2-} \\ s & 2s & s \end{array}$$

$K_{sp} = 2.4 \times 10^{-12} = 4\ s^3$

$s = 8.44 \times 10^{-5}$

$[Ag^+] = 2\ s = 1.69 \times 10^{-4}$

$$E_{cell} = 0.463\ V - \frac{0.0592}{2} \log \frac{Cu^{2+}}{[Ag^+]^2}$$

$$E_{cell} = 0.463\ V - \frac{0.0592}{2} \log \frac{(0.10)}{(1.69 \times 10^{-4})^2}$$

$E_{cell} = 0.269\ V$

89. $Cu^{2+} + 2\ e^- \rightarrow Cu\ (s)$ cathode

$Cu\ (s) \rightarrow Cu^{2+} + 2\ e^-$ anode

$$E_{cell} = E^\circ_{cell} - \frac{0.0592}{2} \log \frac{[Cu^{2+}]\ \text{anode}}{[Cu^{2+}]\ \text{cathode}}$$

$$E_{cell} = 0.00 - \frac{0.0592}{2} \log \frac{0.025}{1.50}$$

$E_{cell} = 0.0526\ V$

As the cell operates the concentration of the 1.50 M solution will decrease, and that of the 0.025 M solution will increase, causing the voltage to decrease. When the concentrations of the solutions become equal, E_{cell} will fall to zero.

91. anode $\qquad 2\ H_2O \rightarrow 4\ H^+ + O_2 + 4\ e^-$

cathode $\qquad \{2\ H_2O + 2\ e^- \rightarrow H_2 + 2\ OH^-\} \times 2$

$\qquad 2\ H_2O \rightarrow 2\ H_2\ (g) + O_2\ (g)$

The H^+ (aq) produced in the anode compartment and the OH^- (aq) in the cathode compartment are produced in exactly equal molar amounts, regardless of the concentration of Na_2SO_4, the electrolysis time, or the current used. When the two solutions are mixed, the result is simply Na_2SO_4 (aq) at its characteristic pH (about 7).

Chapter 21

The p-Block Elements

Exercise

21.1 $2\ BCl_3\ (g) + 3\ H_2\ (g) \rightarrow 2\ B\ (s) + 6\ HCl\ (g)$

Conceptual Exercise

21.1 $2\ Br^-\ (aq) \rightarrow Br_2\ (l) + 2\ e^-$ $\qquad E^\circ_{ox} = -E^\circ_{red} = -1.065\ V$

$SO_4^{2-}\ (aq) + 4\ H^+\ (aq) + 2\ e^- \rightarrow 2\ H_2O + SO_2\ (g)$ $\qquad E^\circ_{red} = 0.17\ V$

$E^\circ_{cell} = -0.90\ V$

$\Delta G^\circ = -nFE^\circ_{cell}$

$$\Delta G^\circ = -2\ \text{mol e}^- \times \frac{96485\ \text{C}}{\text{mol e}^-} \times -0.90\ \text{V} \times \frac{\text{J}}{\text{C V}} \times \frac{\text{kJ}}{1000\ \text{J}} = 174\ \text{kJ}$$

The positive ΔG° indicates that the reaction is nonspontaneous, just as was concluded in Conceptual Example 21.1. The ΔG° values do not agree because the states of the reactants and products differ, for example, H_2SO_4 (l) and Br_2 (g) in Conceptual Example 21.1 and Br_2 (l) and $[H^+]$ = 1M here.

Review Questions

2. Boron (in Group 3A) and silicon (in Group 4A) exhibit a diagonal relationship.

3. Oxygen, silicon, and aluminum are the three most abundant elements in Earth's crust.

4. Aluminum is the most active metal and fluorine is the most active non-metal in the *p* block of elements.

21. a. silver azide b. potassium thiocyanate
c. astatine oxide d. telluric acid

22. a. H_2SeO_4 b. H_2Te c. $Pb(N_3)_2$ d. AgAt

23. bauxite Al_2O_3, boric oxide B_2O_3, corundum Al_2O_3, cyanogen C_2N_2, hydrazine NH_2NH_2, silica SiO_2

24. apatite-Ca, P,O; borax-Na, B,O; corundum-Al, O; quartz-Si, O.

Problems

25. $3\ Mg\ (s) + B_2O_3\ (s) \xrightarrow{\Delta} 2\ B\ (s) + 3\ MgO\ (s)$

27. BH_3 is an electron-deficient structure that does not exist as a stable molecule (the stable species is diborane, B_2H_6). The possibility of resonance structures with B-to-F double bonds lead to a resonance hybrid for BF_3 that is a stable molecule.

29. $Al_2O_3\ (s) + 3\ H_2SO_4 + 15\ H_2O \xrightarrow{\Delta} Al_2(SO_4)_3 \cdot 18\ H_2O$

31. The first key feature is that Al_2O_3 is amphoteric. By adding base Al_2O_3 can be separated from the impurity Fe_2O_3, which is not amphoteric. The second key is the use of molten cryolite, Na_3AlF_6 (l) as a solvent for Al_2O_3 (s). Electrolysis can be conducted at a much lower temperature and in a better electrical conductor.

33. $\Delta H°_{rxn} = \sum \Delta H°_f$ products $- \sum \Delta H°_f$ reactants
$\Delta H°_{rxn} = -1676$ kJ /mol $- (-824.2$ kJ/mol)
$\Delta H°_{rxn} \approx -852$ kJ /mol
This result is only approximate because it is based on data at 298 K, whereas the reaction occurs at a very high temperature. Also, the estimate assumes $\Delta H°_f = 0$ for Fe (s), even though at the temperature of the reaction the stable form of iron is Fe(l).

35. Aluminum could react with strongly acidic foods to produce H_2 (g); the metal would become pitted. In a strongly basic medium (oven cleaner), aluminum could react to produce $[Al(OH)_4]^-$, and again the metal would become pitted.

37. $Na_2C_2 + 2 H_2O \longrightarrow 2 NaOH + C_2H_2$

39. $\ddot{S} = C = \ddot{S}$ $:\ddot{C}l - \underset{\underset{:\ddot{C}l:}{|}}{\overset{\overset{:\ddot{C}l:}{|}}{C}} - \ddot{C}l:$ $:N \equiv C - C \equiv N:$

41. $\left[:\ddot{O} - \underset{\underset{:\ddot{O}:}{|}}{\overset{\overset{:\ddot{O}:}{|}}{Si}} - \ddot{O} - \underset{\underset{:\ddot{O}:}{|}}{\overset{\overset{:\ddot{O}:}{|}}{Si}} - \ddot{O}: \right]^{6-}$ This would be two tetrahedra with a common corner.

43. $\underset{Mg_3(Si_2O_5)(OH)_4}{\overset{\overset{-2 \quad -1}{+2 \ +4-2 \ -2+1}}{}}$ $3(+2) + (-2) + 4(-1) = 0$

45. a. $Si(CH_3)_4$ b. $SiCl_2(CH_3)_2$ c. $SiH(C_2H_5)_3$

47. $PbO_2 + 4 H^+ + 2 e^- \rightarrow Pb^{2+} + 2 H_2O$ $E°_{red} = 1.455$ V
    a. $ClO_3^- + H_2O \rightarrow 2 e^- + 2 H^+ + ClO_4^-$ $E°_{ox} = -E°_{red} = -1.19$ V
    Yes

    b. $H_2O \rightarrow H_2O_2 + 2 H^+ + 2 e^-$ $E°_{ox} = -E°_{red} = -1.763$ V
    No

    c. $Ag^+ \rightarrow Ag^{2+} + e^-$ $E°_{ox} = -E°_{red} = -1.98$ V
    No

    d. $Sn^{2+} \rightarrow Sn^{4+} + 2 e^-$ $E°_{ox} = -E°_{red} = -0.154$ V
    Yes

49. a. Sn (s) $+ 2 HCl$ (aq) $\rightarrow SnCl_2$ (aq) $+ H_2$ (g)
    b. $SnCl_2 + Cl_2$ (g) $\rightarrow SnCl_4$
    c. $SnCl_4$ (aq) $+ 4 NH_3$ (aq) $+ 2 H_2O \rightarrow SnO_2$ (g) $+ 4 NH_4^+$ (aq) $+ 4Cl^-$ (aq)

51. $(CH_3)_2NNH_2 + 4 O_2 \rightarrow 2 CO_2 + 4 H_2O + N_2$

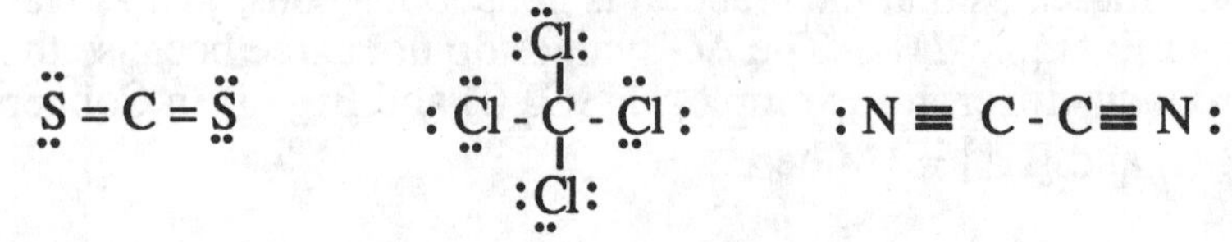

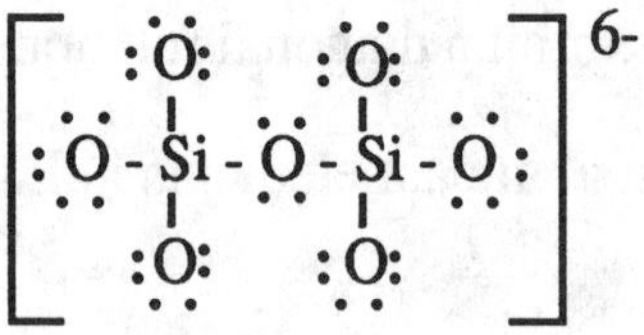

53. a. $NH_2NH_2 + 2\ HCl \rightarrow NH_3NH_3^{2+} + 2\ Cl^-$
b. $3\ Cu\ (s) + 8\ H^+ + 2\ NO_3^- \rightarrow 3\ Cu^{2+} + 2\ NO + 4\ H_2O$
c. $2\ NO\ (g) + O_2\ (g) \rightarrow 2\ NO_2\ (g)$

55. $4(Fe^{3+} + e^- \rightarrow Fe^{2+})$ $E^\circ_{red} = 0.771\ V$
$NH_2NH_3^+ \rightarrow N_2 + 5\ H^+ + 4\ e^-$ $E^\circ_{ox} = -E^\circ_{red} = -(-0.23\ V)$

$4\ Fe^{3+} + NH_2NH_3^+ \rightarrow N_2 + 4\ Fe^{2+} + 5\ H^+$ $E^\circ_{cell} = 1.00\ V$

57. The principal allotropes of phosphorus are white P and red P, with white phosphorus being the more reactive. The molecular structure of white phosphorus consists of individual P_4 tetrahedra. In red phosphorus, the P_4 tetrahedra are joined into long chains.

59. $P_4\ (s) + 3\ KOH\ (aq) + 3\ H_2O \rightarrow 3\ KH_2PO_2\ (aq) + PH_3\ (g)$

61. reduction: $2\ SO_3^{2-} + 3\ H_2O + 4\ e^- \rightarrow S_2O_3^{2-} + 6\ OH^-$
oxidation: $2\ S\ (s) + 6\ OH^- \rightarrow S_2O_3^{2-} + 3\ H_2O + 4\ e^-$

$2\ SO_3^{2-} + 2\ S\ (s) \rightarrow 2\ S_2O_3^{2-} \rightarrow SO_3^{2-} + S\ (s) \rightarrow S_2O_3^{2-}$

63. $Br_2 + 2\ e^- \rightarrow 2\ Br^-$ $E^\circ_{red} = 1.065\ V$
$I_2 + 2\ e^- \rightarrow 2\ I^-$ $E^\circ_{red} = 0.535\ V$
$F_2 + 2\ e^- \rightarrow 2\ F^-$ $E^\circ_{red} = 2.866\ V$
$Cl_2 + 2\ e^- \rightarrow 2\ Cl^-$ $E^\circ_{red} = 1.358\ V$
To displace Br_2 from an aqueous solution of Br^- requires oxidation of Br^-.
$2\ Br^-\ (aq) \rightarrow Br_2\ (l) + 2\ e^-$ $E^\circ_{ox} = -E^\circ_{red} = -1.065\ V$
Only a reduction half-reaction with $E^\circ_{red} > 1.065$ will work, and this must be $Cl_2\ (g) + 2\ e^- \rightarrow 2\ Cl^-\ (aq)$ $E^\circ_{red} = 1.358\ V$. I_2 is too poor an oxidizing agent to work, and I^-, Cl^-, and F^- can only be reducing agents, not oxidizing agents.

65. $I^- + 3\ Cl_2 + 3\ H_2O \rightarrow IO_3^- + 6\ Cl^- + 6\ H^+$
$2\ Br^-\ (aq) + Cl_2\ (g) \rightarrow Br_2\ (l) + 2\ Cl^-\ (aq)$
When the products of the reaction are treated with CS_2 (l), the Br_2 dissolves and the other products remain in the aqueous solution.

67. ICl is an interhalogen compound.
$(CN)_2$ is a pseudohalogen.
NaCl is a halide salt (neither an interhalogen nor pseudohalogen).

69. a. $Cl_2 + H_2O\ (l) \rightleftharpoons HOCl + H^+ + Cl^-$
b. $Cl_2 + 2\ NaOH \rightarrow NaOCl + NaCl + H_2O$
c. $3\ Cl_2 + 6NaOH \xrightarrow{\Delta} 5\ NaCl + NaClO_3 + 3\ H_2O$

71. a. 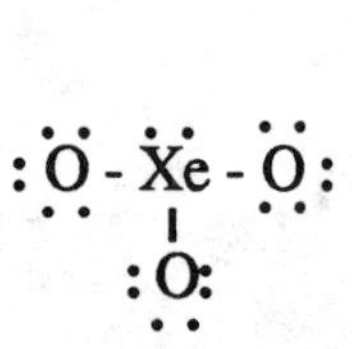

3-bonded atoms
1 L.P.
AXE_3
trigonal pyramidal

b. 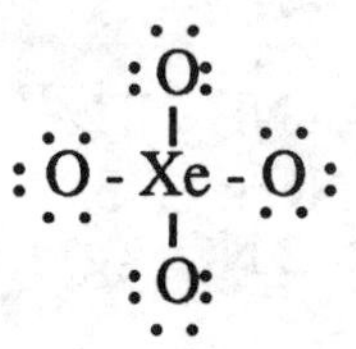

4-bonded atoms
AX_4
tetrahedral

73. a. 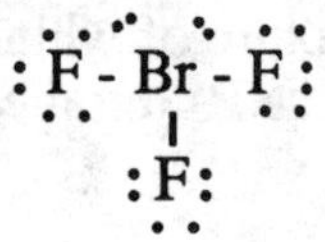

3-bonded atoms
2 L.P.
AX_3E_2
T-shape

b. 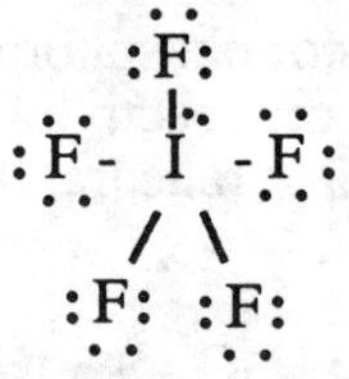

5-bonded atoms
1 L.P.
AX_5E
square pyramidal

75. $Na_2B_4O_7 \cdot 10H_2O + 6\ CaF_2 + 8\ H_2SO_4 \rightarrow 2\ NaHSO_4 + 6\ CaSO_4 + 17\ H_2O + 4\ BF_3$

77. $MnF_6^{2-} + 2\ SbF_5 \rightarrow MnF_4 + 2\ SbF_6^-$
$MnF_4 \rightarrow MnF_2 + F_2\ (g)$

79. Sulfur will melt at 119 °C. Water is heated under pressure to produce steam at temperatures greater than 119 °C. The super-heated steam melts the sulfur underground. Sulfur neither reacts with hot water nor dissolves in it, so the liquid sulfur can be brought to the surface in a pure condition by compressed air.

81. $$2.50\ \text{L} \times 1.75\ \text{M}\ Na_2SO_3 \times \frac{2\ \text{mol}\ H_2S}{\text{mol}\ SO_3^{2-}} \times \frac{100\ \text{mol air}}{1.5\ \text{mol}\ H_2S} \times \frac{\dfrac{62.4\ \text{mm Hg L}}{\text{K mol}} \times 298\ \text{K}}{755\ \text{mm Hg}} =$$
1.4×10^4 L air

83. $Sn^{4+}(aq) \rightarrow 2e^- + Sn^{2+}(aq)$ $\qquad E°_{red} = 0.154$ V
$Sn(s) \rightarrow 2e^- + Sn^{2+}(aq)$ $\qquad E°_{ox} = -E°_{red} = -(-0.137)$ V

$Sn(s) + Sn^{4+}(aq) \rightarrow 2\ Sn^{2+}(aq)$ $\qquad E°_{cell} = 0.291$ V

The reduction of Sn^{4+} by Sn (s) is a spontaneous reaction. Thus, all the tin ion in solution is Sn^{2+} (aq) as long as some solid tin remains.

85. The cell is 335 pm on each side.
$$V = l^3 = (335\ \text{pm})^3 \times \left(\frac{10^{-12}\ \text{m}}{\text{pm}}\right)^3 \times \left(\frac{\text{cm}}{10^{-2}\ \text{m}}\right)^3 = 3.760 \times 10^{-23}\ \text{cm}^3$$
One atom in the unit cell.
$$d = \frac{209\ \text{u}}{3.760 \times 10^{-23}\ \text{cm}^3} \times \frac{1.661 \times 10^{-24}\ \text{g}}{\text{u}} = 9.23\ \text{g/cm}^3$$

Chapter 22

The d-Block Elements

Exercises

22.1 $TiO_2\ (s) + 2\ C\ (s) + 2\ Cl_2\ (g) \xrightarrow{800^\circ C} TiCl_4\ (g) + 2\ CO\ (g)$

$TiCl_4\ (g) + O_2 \xrightarrow{1200^\circ C} TiO_2\ (g) + 2\ Cl_2\ (g)$

22.2 $MnO_4^-\ (aq) + 2\ H_2O + 3\ e^- \rightarrow MnO_2\ (s) + 4\ OH^-\ (aq)$ $E^\circ_{red} = 0.60$ V
MnO_4^- (aq) in basic solution will oxidize any species for which $E^\circ_{ox} > -0.60$ V. This includes Br^- (aq) to BrO_3^- (aq), Br_2 (l) to BrO^- (aq), Ag (s) to Ag_2O (s), NO_2^- (aq) to NO_3^- (aq), S (s) to SO_3^{2-} (aq), and so on.

Conceptual Exercise

22.1 $4\ FeCr_2O_4 + 16\ NaOH + 7\ O_2 \rightarrow 8\ Na_2CrO_4 + 4\ Fe(OH)_3 + 2\ H_2O$

Review Questions

3. Cobalt.

4. No, it is a representative element, indium. The 4*d* subshell is filled and the 5*p* subshell is partially filled.

6. Silver is best followed by copper, and then gold.

9. Chromium, zinc, and nickel.

10. Vanadium, chromium, and manganese.

11. Scandium hydroxide, chromium(III) hydroxide, and zinc hydroxide are all amphoteric.

12. Iron, cobalt, and nickel hydroxides are basic.

13. a. Galvanized iron is zinc coated iron.
b. Chromite ore is $FeCr_2O_4$.
c. Basic copper carbonate is $Cu_2(OH)_2CO_3$.

14. a. Ferromanganese is an iron-manganese alloy.
b. Brass is a copper-zinc alloy.
c. Aqua regia is a mixture of HCl (aq) and HNO_3 (aq).

17. Au

18. Co and Ni lose the two s electrons to form 2+ ions. Iron loses the two s electrons and one 3*d* electron to leave a half filled 3*d* subshell, which is an especially stable electron configuration.

19. a. scandium hydroxide b. iron(II) silicate
c. sodium manganate d. osmium pentacarbonyl

20. a. $BaCr_2O_7$ b. CrO_3 c. Hg_2Br_2 d. Na_3VO_4

Problems

25. In both cases electrons are lost to produce the electron configuration of Ar. With calcium this means the two 4*s* electrons, and the ion Ca^{2+} is formed. With scandium, the 3*d* electron is lost as well, producing Sc^{3+}.

27. (b) The transition elements are metals (EN < 2), but they are not as active as the alkali metals (some of which have EN <1).

29. a. The Ca atom is smaller than the K atom because it has a higher nuclear charge (+20 compared to +19) but the same number of electrons in its noble gas core (18), coupled with the fact that the two 4*s* electrons are not effective in screening one another.
b. The Mn atom is smaller than the Ca atom because it has a higher nuclear charge (+25 compared to +20) and the same number of valence electrons in the same configuration ($4s^2$).
c. The Fe and Mn atoms are about the same size because they have about the same nuclear charge (+25 and +26, respectively), the same number of valence electrons ($4s^2$), and inner shell electrons that are about equally effective in shielding the valence electrons from the nucleus.

31. a. Ti - [Ar] 3*d* [↑][↑][ ][ ][ ] 4*s* [↑↓]

b. Ag - [Kr] 4*d* [↑↓][↑↓][↑↓][↑↓][↑↓] 5*s* [↑] Ag is anomalous because of the stability of a filled *d* subshell.

c. Cr^{2+} - [Ar] 3*d* [↑][↑][↑][↑][ ] 4*s* [ ]

d. Mn^{2+} - [Ar] 3*d* [↑][↑][↑][↑][↑] 4*s* [ ]

33.

+2	+4	+5	+5
V^{2+}	VO^{2+}	VO_2^+	VO_4^{3-}

35. a. $2\ Sc\ (s) + 6\ HCl\ (aq) \rightarrow 2\ ScCl_3\ (aq) + 3\ H_2\ (g)$
b. $Sc(OH)_3\ (s) + 3\ HCl\ (aq) \rightarrow ScCl_3\ (aq) + 3\ H_2O\ (l)$
c. $Sc(OH)_3\ (s) + 3\ Na^+\ (aq) + 3\ OH^-\ (aq) \rightarrow [Sc(OH)_6]^{3-}\ (aq) + 3\ Na^+\ (aq)$
d. $2\ ScCl_3\ (l) \xrightarrow{\text{electrolysis}} 2\ Sc\ (l) + 3\ Cl_2\ (g)$

37. a. $Ba (s) + 2 H^+ (aq) + 2 Cl^- (aq) \rightarrow Ba^{2+} (aq) + H_2 (g) + 2 Cl^- (aq)$; followed by, $Ba^{2+} (aq) + 2 Cl^- (aq) + 2 K^+ (aq) + CrO_4 (aq) \rightarrow BaCrO_4 (s) + 2 K^+ (aq) + 2 Cl^- (aq)$

b. Combine the reduction half-reaction $MnO_4^- (aq) + 2 H_2O + 3 e^- \rightarrow MnO_2 (s) + 4 OH^- (aq)$, $E^\circ_{red} = 0.60$ V with an oxidation half-reaction for which $E^\circ_{ox} > -0.60$ V. For example, $3 NO_2^- (aq) + 2 MnO_4^- (aq) + H_2O \rightarrow 3 NO_3^- (aq) + 2 MnO_2 (s) + 2OH^- (aq)$, $E^\circ_{cell} = 0.59$ V

39. $Mn^{2+} (aq) + 4 H_2O (l) \rightarrow MnO_4^- (aq) + 8 H^+ (aq) + 5 e^-$
$BiO_3^- (aq) + 6 H^+ (aq) + 2 e^- \rightarrow Bi^{3+} (aq) + 3 H_2O (l)$
$2 Mn^{2+} (aq) + 5 BiO_3^- (aq) + 14 H^+ (aq) \rightarrow 2 MnO_4^- (aq) + 5 Bi^{3+} (aq) + 7 H_2O (l)$

41. $[CrO_4^{2-}]$ in the reversible reaction

$$Cr_2O_7^{2-} + H_2O <=> 2 CrO_4^{2-} + 2 H^+$$

is large enough that K_{sp} of $PbCrO_4 (s)$ is exceeded.

43. Fe (s) is the best reducing agent. The reducing agent is oxidized, and $E^\circ_{ox} = 0.440$ V for the half reaction: $Fe (s) \rightarrow Fe^{2+} + 2 e^-$. For the oxidation of $Co (s) \rightarrow Co^{2+} + 2 e^-$, $E^\circ_{ox} = 0.277$ V, and for the oxidation of $Fe^{2+} (aq) \rightarrow Fe^{3+} + e^-$, $E^\circ_{ox} = -0.771$. $Co^{3+}(aq)$ is an oxidizing agent (reduced to $Co^{2+} (aq)$), not a reducing agent.

45. $Fe_2O_3 (s) + 10 OH^- (aq) \rightarrow 2 FeO_4^{2-} (aq) + 5 H_2O + 6 e^-$
$\underline{3(Cl_2 (g) + 2 e^- \rightarrow 2 Cl^- (aq))}$
$Fe_2O_3 (s) + 3 Cl_2 (g) + 10 OH^- (aq) \rightarrow 6 Cl^- (aq) + 2 FeO_4^{2-} (aq) + 5 H_2O$

47. a. $Cu (s) \rightarrow Cu^{2+} + 2 e^-$
$\underline{SO_4^{2-} + 4 H^+ + 2 e^- \rightarrow SO_2 (g) + 2 H_2O}$
$Cu (s) + SO_4^{2-} + 4 H^+ \rightarrow SO_2 + 2 H_2O + Cu^{2+}$ net ionic equation

b. $NO_3^- (aq) + 4 H^+ + 3 e^- \rightarrow NO (g) + 2 H_2O$
$3 Cu (s) + 2 NO_3^- (aq) + 8 H^+ \rightarrow 2 NO (g) + 4 H_2O + 3 Cu^{2+} (aq)$ net ionic equation

49. $NO_3^- + 4 H^+ + 3 e^- \rightarrow NO (g) + 2 H_2O$ $E^\circ_{red} = 0.956$ V
$Ag \rightarrow Ag^+ + e^-$ $E^\circ_{ox} = -E^\circ_{red} = -0.800$ V
E°_{cell} for $NO_3^- + Ag$ is 0.156 V, the Ag reaction will go.
$Au \rightarrow Au^{3+} + 3 e^-$ $E^\circ_{ox} = -E^\circ_{red} = -1.52$ V
E°_{cell} for $NO_3^- + Au$ is -0.56 V, the gold reaction will not occur spontaneously.

51. The Group 1B metals have higher melting points and densities than do the 1A metals. The Group 1B metals do not react with water or mineral acids, whereas the 1A metals do, to liberate $H_2 (g)$. The Group 1B cations form many complex ions, whereas the Group 1A cations form very few. Almost all Group 1A compounds are water soluble, whereas many of the Group 1B compounds are not.

53. a. $3(Hg\ (l) \rightarrow Hg^{2+} + 2\ e^-)$
$\underline{2(NO_3^- + 4\ H^+ + 3\ e^- \rightarrow NO\ (g) + 2\ H_2O)}$
$3\ Hg\ (l) + 2\ NO_3^- + 8\ H^+ \rightarrow 2\ NO + 4\ H_2O + 3\ Hg^{2+}$
b. $ZnO\ (s) + 2\ CH_3CO_2H \rightarrow Zn^{2+} + 2\ CH_3CO_2^- + H_2O$

55. Commercially important elements are generally found in readily available mineral forms (ores) from which they can be extracted by straightforward chemical reactions. Some elements, even though abundant overall in Earth's crust, are too widely scattered in their mineral forms and/or too difficult to extract to make them commercially important.

57. CO molecules have all electrons paired, but the Mn atom does not (Z = 25). $Mn(CO)_5$ is a species with an odd number of electrons and thus, an unpaired electron. The dimer $Mn_2(CO)_{10}$ has all electrons paired.

59. $2(Cu\ (s) \rightarrow Cu^{2+} + 2\ e^-)$
$O_2\ (g) + 2\ H_2O\ (l) + 4\ e^- \rightarrow 4\ OH^-\ (aq)$
$CO_2\ (g) + H_2O\ (l) \rightarrow 2\ H^+\ (aq) + CO_3^{2-}\ (aq)$
$2Cu\ (s) + O_2\ (g) + 3\ H_2O\ (l) + CO_2\ (g) \rightarrow Cu_2(OH)_2CO_3\ (s) + 2\ H_2O\ (l)$
$2Cu\ (s) + O_2\ (g) + H_2O\ (l) + CO_2\ (g) \rightarrow Cu_2(OH)_2CO_3\ (s)$

61. Ag^{2+} is very easily reduced to Ag^+ ($E°_{red} = 1.98$ V), which means that Ag^{2+} is a powerful oxidizing agent. In the hypothetical compounds $AgBr_2$, $AgCl_2$, and AgS, the Ag^{2+} ion would oxidize the anions to the free elements. Thus, the compounds do not exist at all. Oxidation of O^{2-} and F^- to the free elements is much more difficult to accomplish, and so AgO and AgF_2 can be isolated.

63. $Sn^{2+} \rightarrow Sn^{4+} + 2\ e^-$ $\quad E°_{ox} = -0.154$ V
$V^{2+} \rightarrow V^{3+} + e$ $\quad E°_{ox} = 0.255$ V
$Fe^{2+} \rightarrow Fe^{3+} + e^-$ $\quad E°_{ox} = -0.771$ V

a. $MnO_2\ (s) + 4\ H^+ + 2\ e^- \rightarrow Mn^{2+} + 2\ H_2O$ $\quad E°_{red} = 1.23$ V
Sn^{2+}, V^{2+} and Fe^{2+} are all capable.
b. $I_2 + 2\ e^- \rightarrow 2\ I^-$ $\quad E°_{red} = 0.535$ V
V^{2+} and Sn^{2+} are capable.
c. $2\ H^+ + 2\ e^- \rightarrow H_2\ (g)$ $\quad E°_{red} = 0.0$ V
only V^{2+} is capable.

65. $CrO_3 + 6\ H^+ + 6\ e^- \rightarrow Cr\ (s) + 3\ H_2O$

$$2.57\ \text{A} \times 3.25\ \text{h} \times \frac{3600\ \text{s}}{\text{h}} \times \frac{\text{C}}{\text{As}} \times \frac{\text{mol e}^-}{96485\ \text{C}} \times \frac{\text{mol Cr}}{6\ \text{mol e}^-} \times \frac{52.00\ \text{g Cr}}{\text{mol Cr}} = 2.70\ \text{g Cr}$$

67. $K_{eq} = 3.2 \times 10^{14} = \dfrac{[Cr_2O_7^{2-}]}{[Cr_2O_4^{2-}]^2[H^+]^2}$

$$2\,Cr_2O_4^{2-} + 2\,H^+ \Leftrightarrow Cr_2O_7^{2-} + H_2O$$

eq $\quad 1 - 2X \qquad 10^{-9} \qquad X$

$$3.2 \times 10^{14} = \frac{X}{(1 - 2X)^2(10^{-9})^2}$$

assume $X << 1$

$X = 3.2 \times 10^{14} \times 1 \times 10^{-18} = 3.2 \times 10^{-4}$

$[Cr_2O_7^{2-}] = 3.2 \times 10^{-4}$ M which is practically zero when compared to the total Cr(VI) concentration of 1 M.

Chapter 23

Complex Ions and Coordination Compounds

Exercises

23.1 a. Coordination number = 6

$ox\#_{Co} + ox\#_{SO_4^{2-}}$ = species charge

$ox\#_{Co}$ + (-2) = +1

$ox\#_{Co}$ = +3

b. Coordination number = 6

$ox\#_{Fe}$ + 6 x $ox\#_{CN}$ = species charge

$ox\#_{Co}$ + 6 x (-1) = -4

$ox\#_{Fe}$= +2

23.2 a. diamminesilver(I) ion

b. tetrachloroaurate(III) ion

c. pentaamminebromocobalt(III) bromide

23.3 a. $[Co(en)_3]^{+3}$ b. $[CrCl_4(NH_3)_2]^-$ c. $[PtCl_2(en)_2]SO_4$

23.4 a. Different structural isomers are not possible.

b. $[Pt(NH_3)_4][CuCl_4]$ is the other isomer.

23.5 a.

For a strong field ligand (CN^-), the electrons are paired in the lower orbitals.

Co^{3+}[Ar]

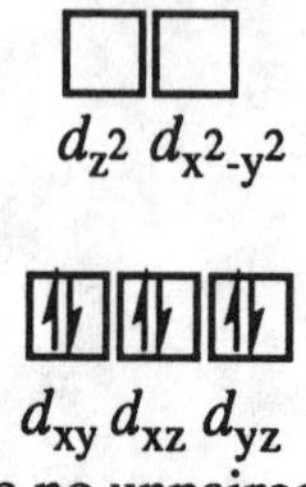

There are no unpaired electrons.

b. Ni^{2+}[Ar] 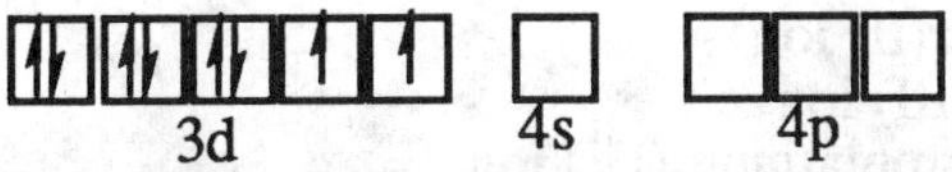

Cl^- is usually a weak field ligand, but the distribution of the eight 3*d* electrons would be the same even if it were strong field. There are 2 unpaired electrons.

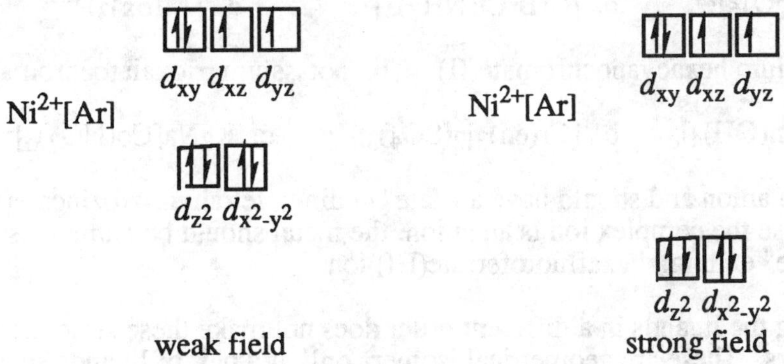

Conceptual Exercise

23.1 Assuming that the ethylenediamine molecule can link only to adjacent coordination sites there is only one form of the molecule and no geometric or optical isomers.

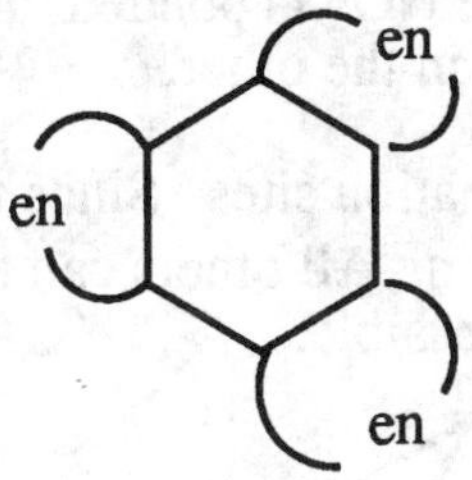

Review Questions

8. Ammonia is a Brønsted-Lowry base because it will accept a H^+ through the lone pair of electrons of the N atom. It is a Lewis base because it can donate this same pair of electrons to a covalent bond.

11. a. water b. ammonia c. ethylenediamine d. oxalate ion

12. a. chloride ion b. carbonate ion c. nitrate ion d. hydrogen sulfite ion

14. The central atom is part of a complex anion.

15. The nitro group is the nitrite ion when bonded to the central atom through lone-pair electrons of the N atom; the nitrito group is bonded through lone-pair electrons on one of the oxygen atoms.

Problems

21. a. 6 b. 4 c. 4 d. 6

23. a. +2 b. +3 c. +3 d. +3

25. a. tetraamminecopper(II) ion
b. hexafluoroferrate(III) ion
c. tetraamminedichloroplatinum(IV) ion
d. tris(ethylenediamine)chromium(III) ion

27. a. $[Fe(H_2O)_6]^{2+}$ b. $[CrBrCl(NH_3)_4]^+$ c. $[Al(ox)_3]^{3-}$

29. a. potassium hexacyanochromate(II) b. potassium trioxalatochromate(III)

31. a. $Na_2[Zn(OH)_4]$ b. $[Cr(en)_3]_2(SO_4)_3$ c. $K_2Na[Co(NO_2)_6]$

33. a. It is an anion and should have an "ate" ending: tetrahydroxozincate(II) ion.
b. Because the complex ion is an anion, the metal should be named last and given an "ate" ending: hexafluoroferrate(III) ion.

35. a. Listing the ligands in a different order does not make these structures isomers. These are geometrical isomers only if a pair of ligands such as Cl, NH_3 or H_2O is cis in one structure and trans in the other, but this cannot be indicated by the formula alone.
b. Yes, these are structural isomers.
c. No, these are not isomers. They contain different numbers of Cl because the oxidation state of Co is +3 in one ion and +2 in the other.
d. Yes, these are structural isomers. The NO_2 is bonded through the N atom in one structure and through the O atom in the other.

37. The en can only bond at adjacent coordination sites. Since the other four sites are all filled by Cl^-, there exists only one form. All others can be made by a rotation of the first. Cis-trans isomerism is not possible.

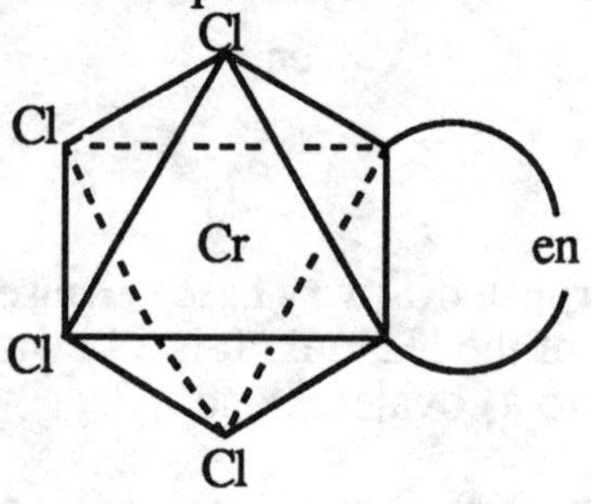

39. Yes.

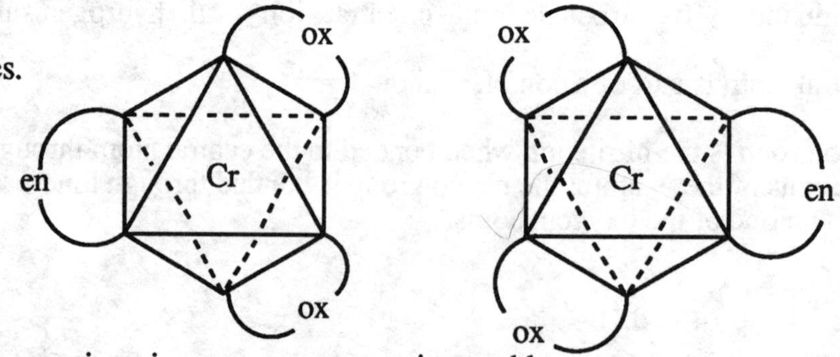

These mirror images are not superimposable.

41. a.

Mn^{2+} [Ar]

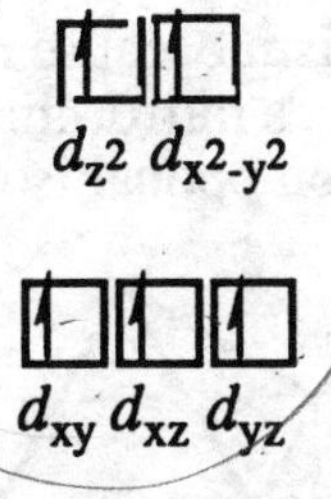

weak field-"high spin"

Mn^{2+} [Ar]

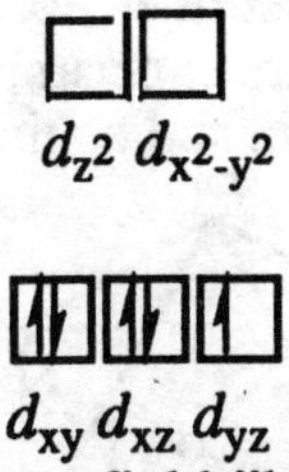

strong field-"low spin"

Paramagnetic. The electron configuration of Mn^{2+} is $[Ar]3d^5$, and because of the odd number of electrons, there must be at least one unpaired electron present. If the complex is of the "high spin" type, the number of unpaired electrons is five; for the "low spin" type it is one.

b.

Fe^{2+} [Ar]

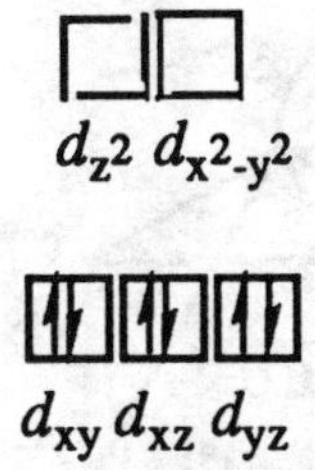

Diamagnetic. The electron configuration of Fe^{2+} is $[Ar]3d^6$. The complex ion should be "low spin" because CN^- is a strong field ligand. The six electrons should all be paired in the lower-energy group of three 3*d* orbitals.

43. $[Cr(H_2O)_6]^{3+}$ is violet and $[Cr(NH_3)_6]^{3+}$ is yellow. NH_3 is a stronger field ligand, causing a greater Δ value, and requiring light of a higher frequency to stimulate an electronic transition. Light of a higher frequency has a shorter wavelength. Violet is a shorter wavelength than yellow. $[Cr(NH_3)_6]^{3+}$ should absorb violet and transmit yellow light.

45. The trans isomer does not exhibit optical isomerism because the structure and its mirror image are superimposable. The cis isomer's mirror image is not superimposable, and so the cis isomer does exhibit optical isomerism.

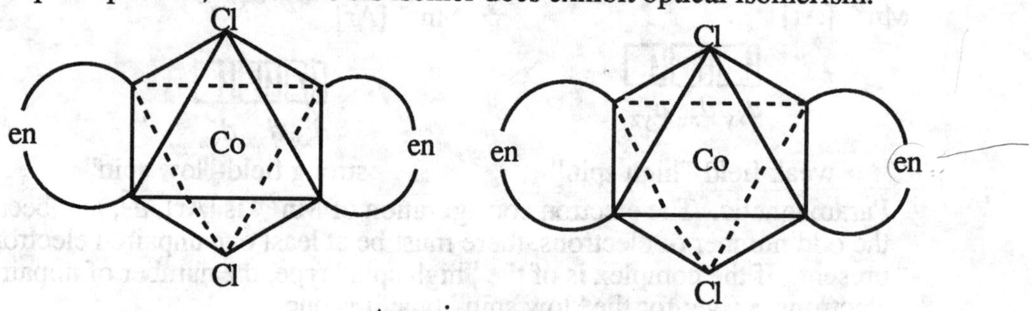

trans isomers

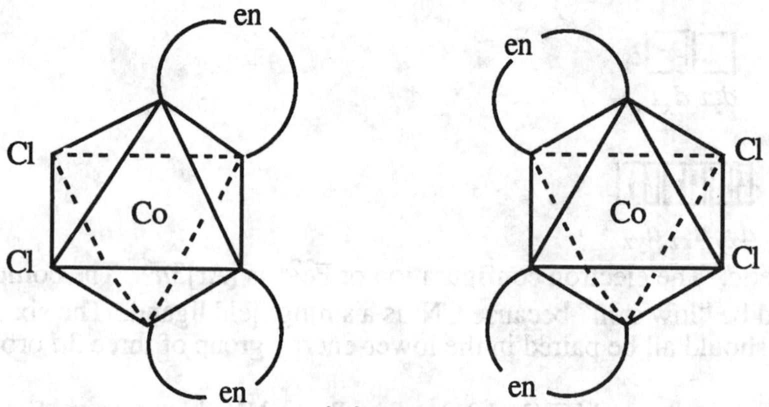

cis isomers

47. $[PtCl_2(NH_3)_2]$ is a square planar complex in which two identical ligands can be either cis or trans. $[ZnCl_2(NH_3)_2]$ is a tetrahedral complex, and does not exhibit cis-trans isomerism.

49. Cr^{3+} has an odd number of electrons so it will be paramagnetic with any ligand. Cr^{3+} has the electron configuration $[Ar]3d^3$. The three 3*d* electrons will remain unpaired in the lower-energy set of three 3*d* orbitals, regardless of whether the ligands (L) are strong field or weak field.

51. Because the complex cation and anion each have Pt^{2+} as the central ion, the formula should be based on a doubling of the empirical formula 2 x $(PtCl_2 \cdot 2NH_3)$ = $Pt_2Cl_4 \cdot 4NH_3$. The true formula then is $[Pt(NH_3)_4][PtCl_4]$.

53.

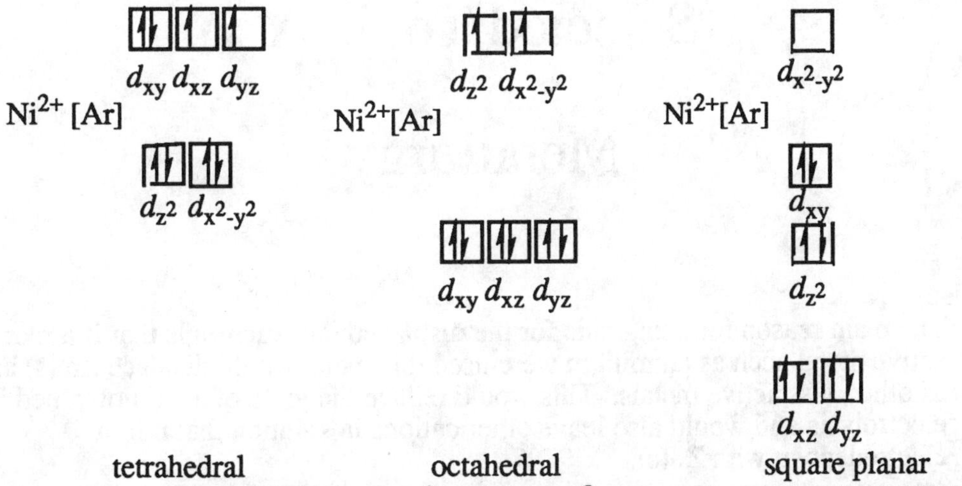

The electron configuration of Ni^{2+} is [Ar]3d^8. The complex ion $[Ni(CN)_4]^{2-}$ is a strong-field complex in which all the electrons are paired (diamagnetic). The assignment of eight electrons to the tetrahedral splitting diagram would leave two electrons unpaired. The assignment of eight electrons to the octahedral splitting diagram would also leave two electrons unpaired. The assignment of eight electrons to the diagram for a square planar complex would fill the four lowest energy *d* orbitals with electron pairs and leave the highest energy *d* orbital empty. The structure of the complex ion is square planar.

Selected Topic A

Metallurgy

Exercise

A.1 The main reason for using zinc for the displacement reaction is that if a more active metal such as aluminum were used this metal would displace Zn (s) as well as other, less active metals. This would reduce the yield of Zn (s) obtained in the electrolysis and would also leave other cations in solution that might electrodeposit with Zn(s).

Review Questions and Problems

17. $2\ Ag_2O(s) \xrightarrow{\Delta} 4\ Ag(s) + O_2(g)$ is the decomposition reaction that occurs when silver oxide is heated.

19. $2\ [Ag(CN)_2]^{-}(aq) + Zn(s) \longrightarrow 2\ Ag(s) + [Zn(CN)_4]^{2-}(aq)$ summarizes the displacement of silver from its cyano complex by zinc.

20. $2\ FeTiO_3(s) + 6\ C(s) + 7\ Cl_2(g) \xrightarrow{\Delta} 2\ FeCl_3(s) + 6\ CO(g) + 2\ TiCl_4(g)$ is a plausible equation for the formation of $TiCl_4(g)$ from ilmenite.

22. Dissolving: $ZnO(s) + H_2SO_4(aq) \longrightarrow ZnSO_4(aq) + H_2O(l)$

Displacement: $Zn(s) + CdSO_4(aq) \longrightarrow Cd(s) + ZnSO_4(aq)$

Electrolysis: $2\ ZnSO_4(aq) + 2\ H_2O(l) \xrightarrow{\text{electrolysis}} 2\ Zn(s) + 2\ H_2SO_4(aq) + O_2(g)$

The $H_2SO_4(aq)$ solution is all that remains after electrolysis, and it then can be reused to dissolve more ZnO(s).

ADDITIONAL PROBLEMS

23. An ore containing a high percentage of a metal may not be the best industrial source of the metal because the metal may be in a compound that is chemically difficult to reduce. For example, many clays contain high percentages of aluminum, but the metal is difficult to extract from them. Another difficulty can be the by-products produced during the extraction process. If these by-products complicate the purification process or are hazardous or difficult to dispose of, their formation will make the use of the ore uneconomical.

25. Iron obtained by direct reduction has a lower carbon content than blast furnace iron because blast furnace iron was reduced with elemental carbon, while DR iron was reduced with a mixture of CO(g) and $H_2(g)$; there was not as much carbon used in the reduction process. Also, the iron never melts in the DR process and is therefore less likely to become contaminated with carbon.

27. The metallurgy of cobalt should more closely resemble that of iron than that of titanium, since iron and cobalt are in the same grouping of the periodic table (the iron triad) and have many properties in common, whereas titanium is at almost the other end of the transition series.

29. Chlorination: $Sn(s) + 2\,Cl_2(g) \longrightarrow SnCl_4(l)$

Hydrolysis: $SnCl_4(l) + (2+x)H_2O(l) \longrightarrow SnO_2{\cdot}xH_2O(s) + 4\,HCl(aq)$

Dehydration: $SnO_2{\cdot}xH_2O(s) \xrightarrow{\Delta} SnO_2(s) + x\,H_2O(g)$

Reduction: $SnO_2(s) + 2\,C(s) \xrightarrow{\Delta} Sn(l) + 2\,CO(g)$

31. $$\text{mass of ore} = 1 \times 10^7 \text{ kg pig iron} \times \frac{95 \text{ kg Fe}}{100 \text{ kg pig iron}} \times \frac{1 \text{ kmol Fe}}{55.85 \text{ kg Fe}} \times \frac{1 \text{ kmol } Fe_2O_3 = \text{hematite}}{2 \text{ kmol Fe}} \times \frac{159.7 \text{ kg } Fe_2O_3}{1 \text{ kmol } Fe_2O_3} \times \frac{100 \text{ kg ore}}{82 \text{ kg hematite}} =$$

1.7×10^7 kg ore.

33. Determine $[Cr_2O_7^{2-}]$ in the 10.00-mL aliquot.

$$\text{moles of } Cr_2O_7^{2-} = 0.1387 \text{ g } BaCrO_4 \times \frac{1 \text{ mol } BaCrO_4}{253.34 \text{ g } BaCrO_4} \times \frac{1 \text{ mol } CrO_4^{2-}}{1 \text{ mol } BaCrO_4} \times \frac{1 \text{ mol } Cr_2O_7^{2-}}{2 \text{ mol } CrO_4^{2}} = 2.737 \text{ x } 10^{-4}$$

$$[Cr_2O_7^{2-}] = \frac{2.737 \text{ x } 10^{-4} \text{ mol } Cr_2O_7^{2-}}{10.00 \text{ mL} \times \dfrac{1 \text{ L}}{1000 \text{ mL}}} = 0.02737 \text{ M}$$

Then determine the mass of Cr in the 250.0 mL of solution.

$$\text{Cr mass} = 250.0 \text{ mL} \times \frac{1 \text{ L}}{1000 \text{ mL}} \times \frac{0.02737 \text{ mol } Cr_2O_7^{2-}}{1 \text{ L}} \times \frac{2 \text{ mol Cr}}{1 \text{ mol } Cr_2O_7^{2-}} \times \frac{51.996 \text{ g Cr}}{1 \text{ mol Cr}} = 0.7116 \text{ g Cr}$$

And finally determine the % Cr in the stainless steel.

$$\% \text{ Cr} = \frac{0.7116 \text{ g Cr}}{5.000 \text{ g steel}} \times 100\% = 14.23\% \text{ Cr}$$

Selected Topic B

Bonding in Metals and Semiconductors

Review Questions and Problems

3. Al is the best electrical conductor; of the four elements listed, it is the only metal. S_8 exists as discrete covalently bonded molecules that permit no electron movement beyond the molecule. As and Ge are both metalloids that do not conduct electricity until quite a large voltage is applied to them.

4. Cu(s) is a metal and a good electrical conductor. A solution of NaCl(aq) contains Na^+(aq) and Cl^-(aq) ions that conduct a current. A solution of CH_3COOH(aq) contains a few H_3O^+(aq) and CH_3COO^-(aq) ions and conducts weakly. But there are no ions in a solution of I_2(s) in CCl_4(l); it is an electrical insulator. It is a nonelectrolyte solution with essentially no electric charge carriers.

15. If the dopant has an extra valence electron, the resulting semiconductor is *n*-type, since negative electrons carry the current. This is the case with (a) Ge doped with As, and (c) Si doped with Sb. If the dopant has one fewer valence electron, the resulting semiconductor is *p*-type, since positive holes carry the current. This is the case with (b) Ge doped with Al and (d) Si doped with B.

16. Elements from periodic family 5A should produce *n*-type semiconductors when added to pure Ge. Of course, their atoms should also be about the same size as Ge atoms so the addition does not disrupt the crystal lattice. P, As, and Sb would be suitable. Elements from periodic family 3A should produce *p*-type semiconductors when added to pure Ge. Those with atoms about the same size as Ge atoms are Al, Ga, and In.

ADDITIONAL PROBLEMS

21. We obtain the number of valence electrons (v.e.), outer-shell electrons, from the electron configuration of each element.

a. [Cu] = [Ar] $3d^{10}\,4s^1$	1 v.e.	[S] = [Ne] $3s^2\,3p^4$	6 v.e.
average = 3.5 v.e.			
b. [Zn] = [Ar] $3d^{10}\,4s^2$	2 v.e.	[Se] = [Ar] $3d^{10}\,4s^2\,4p^4$	6 v.e.
average = 4 v.e.			
c. [Pb] = [Xe] $4f^{14}\,5d^{10}\,6s^2\,6p^2$	4 v.e.	[O] = $1s^2\,2s^2\,2p^4$	6 v.e.
average = 5 v.e.			
d. [Ga] = [Ar] $3d^{10}\,4s^2\,4p^1$	3 v.e	[P] = [Ne] $3s^2\,3p^3$	5 v.e.
average = 4 v.e.			

ZnSe and GaP meet the average number of valence electrons criterion for being semiconductors.

23. A crystal of silicon should be fairly brittle. The bonding is almost exactly like that in diamond, that is, all covalent bonds. There should be little metallic bonding; there is a substantial band gap.